Merry Christmas Emily

December 1979

island forest year

ELK Island National Park

Publication of this book
was assisted by a grant
from Alberta Culture to
the National and Provincial
Parks Association (Alberta).

island forest year ELK Island National Park

written and
illustrated by
Deirdre Griffiths

The
University
of Alberta
Press

1979

First published by
The University of Alberta Press
Edmonton, Alberta, Canada
1979

Canadian Cataloguing in Publication Data

Griffiths, Deirdre, 1939–
Island forest year

Bibliography: p.
Includes index.
ISBN 0-88864-060-9

1. Natural history—Alberta—Elk Island National Park. 2. Elk Island National Park, Alta. I. Title.
QH106.2.A4G75 500.9′7123′3 C79-091043-8

Book design by Jorge Frascara
Jacket illustration by Deirdre Griffiths

Printed by Hignell Printing Limited,
Winnipeg, Manitoba, Canada.

For my Mother and Father
in loving appreciation

I am indebted to naturalists and biologists, past and present, whose painstaking studies have for years enriched my understanding of casual field observations.

With respect to the preparation of this book, I would like to mention those who generously responded to my specific enquiries: Don Bowen of the University of British Columbia, Francis Cook of the National Museum of Canada, Peter Kevan of the University of Colorado, Walter Sheppe of the University of Akron, Chip Wesloh of the Provincial Museum of Alberta, Robert E. Wrigley of the Manitoba Museum of Man and Nature; and David Boag, Don Emerson, Joe Nelson, Ted Pike, Wayne Roberts, and Bob Vance, all of the University of Alberta.

I am particularly grateful to Jack Schick, the Naturalist who succeeded me at Elk Island, for his readiness to discuss many aspects of the book, as well as his helpful comments on the manuscript. Special thanks also go to my mother, and my husband, Graham, for their constructive suggestions during the long period of writing.

D.E.G.

Contents

Illustrations

Back before man was much advanced toward the human condition there was nothing in the world but wild country. Man was a wilderness species by necessity and not by choice until the time came when we developed the skills and tools needed to modify the landscape. Over tens of thousands of years, however, man evolved his social behavior and cultural attributes in wilderness environments, shaped by forces of fire and storm, abundance and scarcity, floods and droughts, predators and prey. The roots of our attitudes, drives, or perhaps they can be called instincts, that govern our activities today lie far back in our wilderness past, inherited perhaps from prehuman ancestors.

If aggression and territoriality are basic human traits, then sympathy and empathy are also human. Man destroys a forest but then feels ill at ease and builds a shrine to the spirits of the wood.

Raymond F. Dasmann
A Different Kind of Country

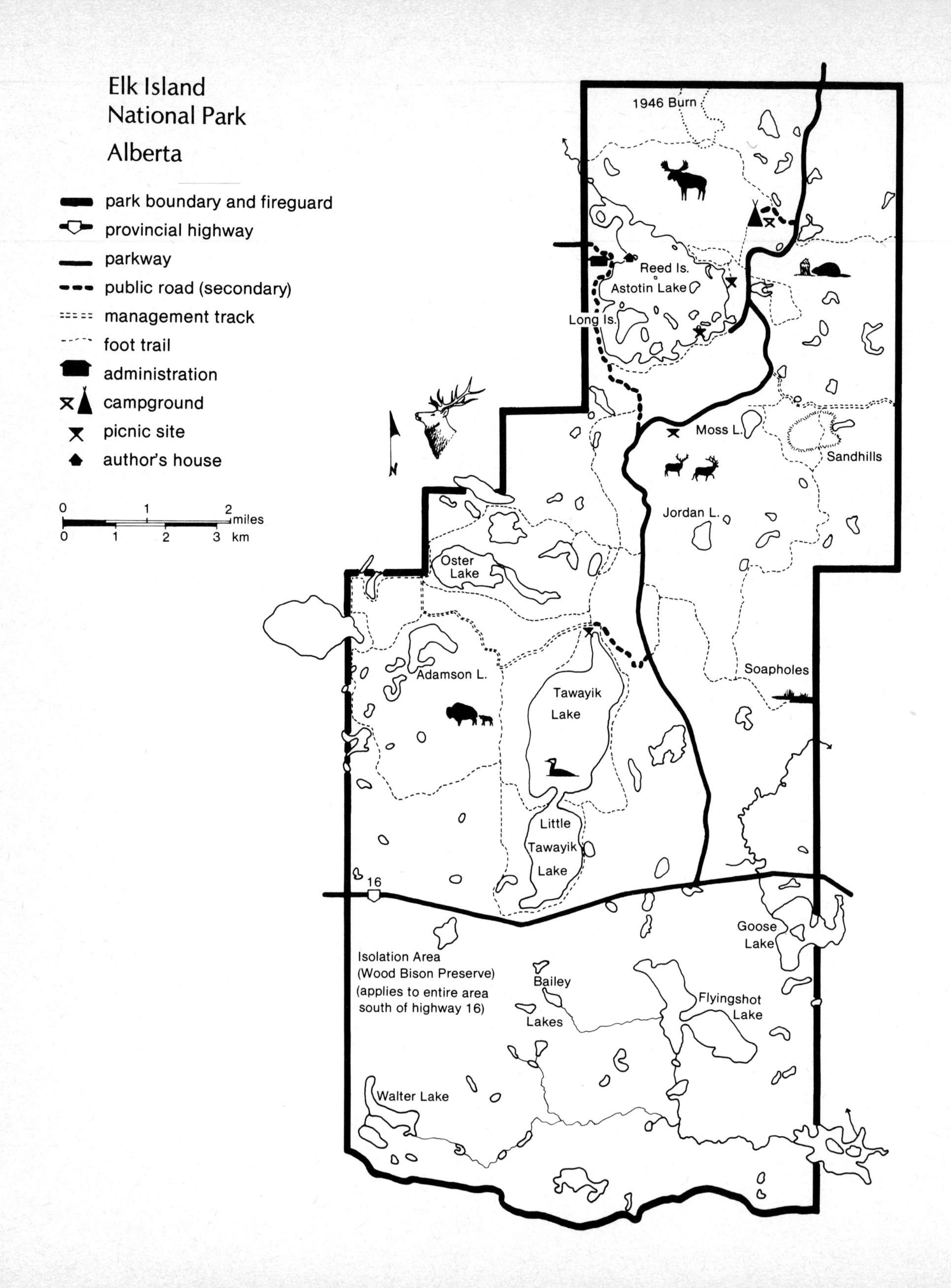

Elk Island
National Park
Alberta
park boundary and fireguard
provincial highway
parkway
public road (secondary)
management track
foot trail
administration
campground
picnic site
author's house
0
1
2
miles
0
1
2
3
km
1946 Burn
Reed Is.
Astotin Lake
Long Is.
Moss L.
Sandhills
Jordan L.
Oster
Lake
Adamson L.
Tawayik
Lake
Soapholes
Little
Tawayik
Lake
16
Goose
Lake
Isolation Area
(Wood Bison Preserve)
(applies to entire area
south of highway 16)
Bailey
Lakes
Flyingshot
Lake
Walter Lake

The setting

Bison, once the foundation of the plains economy, lived only in memory by the time homesteaders began to fan out over the virgin prairies and parklands of central Alberta. The settlers' last "hunt" was to gather up the great beasts' weathered bones, littering the grass in death as the animals had roamed them in life, and sell them to be ground into fertilizer. In the first three decades of the twentieth century farming completely replaced hunting and trapping over the southern half of the province. Grassland that had grown undisturbed for millennia was turned to bare soil for wheat and other crops; fences gradually enclosed and tamed the west's wide open spaces; wagon tracks scored links from homestead to homestead, and from them to the settlements; in time spur railway lines webbed the landscape with shining steel. Many of the old fur trading posts soon crumbled into oblivion, but others like Edmonton and Calgary mushroomed into sprawling frontier towns. And the remnants of the once-proud, nomadic Indians trekked to prescribed reservations where they struggled to cope with the doom that had overtaken them so unexpectedly.

As the settlers pushed east from Edmonton they soon encountered the forested, rolling moraineland of the Beaver Hills. The trees seemed to stretch to infinity in every direction, and were considered a nuisance by most. Fires set to help clear the land often raged out of control and burned deep into the Hills again and again.

Destruction of this wildlife habitat, together with the hunting and trapping that continued until crops and livestock brought dependable returns, resulted in the local extirpation of several mammals—wolf, bear, cougar, otter, marten, and beaver. Others declined to extreme scarcity, including moose, elk, and mule deer.

Fortunately, five local men saw the inevitable outcome of such sustained persecution and took action, petitioning the federal government to establish a fenced elk sanctuary of at least 16 square miles at Astotin Lake in the Beaver Hills. By then elk were extremely scarce everywhere in Alberta. The government responded to the plea, and on March 28, 1906 signed the agreement that created the Game Preserve.

Once the fence was completed, a drive combing the Beaver Hills forced some elk into the Preserve—a count in 1907 tallied 24 elk and 35 mule deer. Additionally, the area harboured a few moose and a variety of small mammals and birds.

Also in 1907 four hundred plains bison arrived, part of the seven hundred destined to stock the new Buffalo National Park near Wainwright, Alberta—a park which was ceded to the Armed Forces in 1947—but which could not be accommodated until the fencing there was finished. That done by 1909, the bison in the area now known as Elk Island National Park were rounded up. The fifty or so who eluded the riders thus became the direct ancestors of today's herd.

The park was enlarged by 36 square miles in 1922 at which time it was obvious that the increasing numbers of moose, elk, and bison needed more range; and by another 24 square miles in 1947. Both additions were fenced, but the smaller block was cut off from the rest of the park by highway 16, and has been used since 1965 to maintain a small herd of pure wood bison isolated from outside influences, particularly disease. During the past ten years brucellosis and tuberculosis have been eradicated and the small herd has grown. Then in 1977 four calves were sent to the Calgary Zoo. In 1978 a transplant was made to northeastern Jasper National Park in the hopes of establishing a wild population there. If this experiment succeeds several more transplants may be made in the next few years to suitable areas in the sub-species' former range.

In 1930 Elk Island became officially a National Park, and is the only one in Canada with a boundary fence. Initially it kept the elk secure; later it prevented the bison from wandering into the adjacent farmland that almost encircles the park. However, a few elk and moose manage to leap over the eight feet of wire stock-fence, and have moved south where parts of the Beaver Hills are still sparsely populated. To other mammals and birds the fence is little or no hindrance, and this is one reason why certain species have not been reintroduced, particularly black bear, cougar, and wolf. The fish could no longer support otters, hence no attempt has been made to re-establish them either.

Lynx have increased again without assistance but apparently are not permanent residents. Their staple prey, the snowshoe hare, never reaches abundance within the park because of competition with the ungulates for winter browse.

After unsuccessful attempts to establish beaver in the early 1940s, they appeared on Astotin Lake as immigrants a decade later. These throve and soon spread throughout the park where they are now abundant. Their dam building has added to the number of permanent ponds, as well as enlarging many existing ones. However, now that the population is high, litter size has decreased. Dispersal, and predation on the young kits also combine to curb population growth. Contagious diseases, like tularemia, can take a heavy toll of life, yet the 1974 outbreak was not a serious one.

Mule deer have declined drastically since the appearance of white-tailed deer in the early 1950s, but the reasons have not been investigated.

Early management policy encouraged large populations of the park's prime public attractions—bison, moose, and elk. As a result all browse was severely hedged, and less resistant plants, especially saskatoon and red-osier dogwood, became scarce. It was even necessary to feed the bison hay in winter. However, during the past decade periodic reductions have brought numbers down from the peaks of approximately 2,000 head of each species to between 350 and 500 head. Nevertheless, it will be many years before the shrubs recover, while management of the park's hoofed mammals will always be difficult and controversial.

At first glance Elk Island seems monotonous—popular "bush" interrupted here and there by small meadows, lakes, ponds, and bogs. The landscape lacks drama to excite the senses and arouse awe. But this superficial sameness is a deceptive cloak, for on longer, more intimate acquaintance one discovers an intriguing diversity that ramifies endlessly. In a lifetime it would be possible to know the park well, never completely. Some of this diversity can be stated numerically—33 species of mammals, 217 of birds, 1 of reptiles (the garter snake), 4 of amphibians, and 4 of fish (annotated species lists are given in Appendices 1 to 5). A comprehensive listing of invertebrates—insects, spiders, harvestmen, mites, crustaceans, molluscs, worms, and countless lesser lives—would be an astronomical undertaking. Catalogued plant species number 308, yet even this is not a complete listing and omits most fungi, algae, lichens, true mosses, and liverworts.

Long-past geographical events also contribute to the present variety and character of the landscape and its life. Massive, ancient beds of shale and sandstone underlie the moraine, but tilt so slightly that they give little direction to surface run-off. Consequently, no well-defined stream pattern has developed. The park's one permanent stream, near the east boundary, meanders lethargically and in dry seasons shrinks to a chain of pools. Other channels flow only during the spring melt and after exceptionally heavy, prolonged rains. When the great continental ice sheet stagnated here several thousand years ago and melted away, it dumped its load of boulders, gravel, and rock flour leaving a hummocky upland of small rounded hills and bowl-like depressions—the geographer's "knob and kettle topography." Water stays trapped in these hollows because much of the moraine is impervious clay. The 220, more or less, ponds and lakes of today are all shallow (Astotin Lake alone exceeds 10 feet in places, but is no deeper than 25 feet), team with aquatic plants and animals, and are crowded with waterfowl.

When the last of the ice vanished, plants were near at hand and soon recolonized the barren ground, first with a low mat of tundra-like growth. Before long, however, trees appeared, pines and poplars dominating the early forests. Later white spruce and balsam fir mingled with them, especially on heavy, moist soils. More recent sequences of plant communities are being deduced from analysis of pollen preserved in bogs. On the basis of present knowledge from this study it seems that about 7,500 years ago a warm, dry period spread open parkland into this region. About 3,500 years later the climate became cooler and wetter again, more like today's. Forest closed in once more, and for the past several hundred years the Beaver Hills have been covered by a mixedwood forest of white spruce, white birch, aspen, and balsam poplar. Black spruce and larch are restricted to bogs. Such was the scene described by early explorers.

The current predominance of poplar is a result of fires and settlement. Each new spruce tree must grow from seed, but poplars recover quickly from moderate fires and clearing by sending up a bristling of shoots from their spreading root systems. In the open, domed groves develop from such regeneration. These are gradually merging and in time will eclipse all the old meadows. However, some steep, south-facing slopes are naturally treeless, being too exposed to extremes of heat and drought.

Jack pine no longer grows on the Sandhills east of Moss Lake, but bearberry, slender blue beardtongue, hairy wild rye, and false dandelion are some of the plants found there exclusively while others—wood lily and smooth aster to name two—are more abundant there. Oddly enough, the blueberry, *Vaccinium myrtilloides*, growing otherwise only in bogs, forms extensive thickets on the Sandhills.

Variety in the wetlands is partly an outcome of time and change. The shallowest basins filled in first and are now sedge and willow fens or black spruce bogs, those deficient in nutrients becoming the bogs. They are not uniform however: some fens have more willows than sedges, other scarcely any willows and perhaps a central pool; bogs may be densely treed, open and park-like, or have a ring of trees around a quaking centre. Pond and lake levels respond quickly to changes in precipitation and beaver activity.

The Soapholes are unique, and were formed by mineral-saturated water surfacing in a broad, shallow swale. There are several pools of liquid mud, each with a central clump of bulrushes. In dry weather much of the alkaline mud cakes superficially and is a popular salt-lick for bison, elk, deer, and moose.

This completes the basic outline of Elk Island National Park and its history. However, for the writer to evolve beyond plain description a place must be experienced day to day through the seasons. Factual observation and recording transcend to empathy when the human becomes an integral element in countless interacting events, many of them fleeting, and fragmentary, and minor, but all facets of a unity. When I arrived in Elk Island at the beginning of 1969 everything was new and strange. By the time I left in June of 1972 the park had become a part of my being, as immediate and complete in retrospect as the indelible places of my childhood.

I had difficulty deciding where to begin this book since the seasons circle through the years without apparent pause. January 1 is the start of our calendar year but has no particular significance otherwise. If indeed there is a beginning to the year it is December 21, the day of the winter solstice, when the sun, so to speak, turns and begins yet again its long northward pilgrimage to summer.

December

The first day of a new year. Over the next six months the days will lengthen, but for a long time yet the increment of light and heat will be almost imperceptible. The nadir of the old year has yet to come.

Yesterday snow fell all day from an amorphous sky; fell softly, windlessly. The flakes heaped up on horizontal branches, in the forks of twigs, on the shelves of bracket fungi, on the tined fingers of the spruce. Tall grasses and sedges yielded to their mounded weight. It was still snowing as the afternoon light faded; and when I stepped out for a few minutes in late evening, I felt the snow's wet coldness on my face, sensed the muffled silence.

During the night the clouds have cleared away. The temperature has dropped to 0 degrees Fahrenheit. I stand out in the predawn darkness under a canopy of pulsating starlight. The air is completely still, yet seems to tingle with unheard sound. Slowly the blackness overhead flushes deep violet as a pallor brightens the east like frost growing across a window in the night. The silhouetted trees gain substance; are again three dimensional. The light spreads over the sky washing out the violet and extinguishing the stars. Now the eastern horizon reddens, the thin livid line fading to peach, to green, to blue above. At last, long after it seems inevitable, the sun itself flares into view, its orange radiance flinging on a cloak of illusory warmth.

But it is light alone that stirs the black-capped chickadees. Five come past, feeding as they move. Chickadees are never quiet or still for long. Their namesake calls, often preceded by a short liquid twitter, are always vivacious. The birds themselves, feathers fluffed against the cold, look like animated feather-puff toys. I follow the movements of one for a minute or so. It suddenly drops from one branch to a lower. Head cocked from side to side, the bird peers at the branch's wrinkled base, then pecks in the crevices

with its tiny black beak. It flits to a twig and, hanging upside down, investigates a cluster of dead leaves. Snow dusts down. Next the chickadee flies to the main trunk. Clinging there it probes the deep fissures of the bark carefully, and apparently finds food, probably some small hibernating insect or spider. A broken branch next attracts its attention, and it examines the frayed end. All this time the chickadee has repeatedly flicked its tail and wings, an action that seems purely reflex. Abruptly then, the bird flies off to another poplar. The other chickadees have been similarly engaged, each seeking food independently. Yet the group is cohesive, and gradually I realize that I am being left behind. As the sounds of the chickadees diminish, I can hear a woodpecker far off, tapping methodically.

Later in the morning I go for a long walk. Although yesterday's storm was not the first snow of the winter, it is not deep enough yet for snowshoes, and I step through the fluff almost unhindered. The air is now a little warmer and the snow squeaks less shrilly.

The trail leads first through maturing poplar forest. Trees and undergrowth are webbed with white. The slanting light, reflecting from and shadowing a multitude of surfaces appears directionless, diffused. The dark masses of the trees are broken up by the snow, and the whole forest is somehow insubstantial. Even the spruce, huddling together at the edge of the sedge meadow, are so covered that only fragments of their night-green foliage are visible. The forest's distance lightens instead of darkening. I am adrift, and feel I could be loosed from the ground at will to go floating between the pillared trunks.

A steady whacking on dead wood brings me back to reality. As I near the source the sound becomes sharper, almost like the bite of an axe. But it originates from high up. Then I see a pileated woodpecker. Nearly as large as a crow, he is all black except for a broad white stripe along the side of his face and neck, white chin, and flaming red crest. As the bird is an adult, the red extends to the beak and a moustache mark. He goes on pounding. When he breaks into the soft inner wood his pace quickens; large chips are flicked aside and sink into the snow below. He stops and probes. A few moments later he shifts his position slightly and starts excavating again. As he moves his claws rasp drily. Finally, fifteen minutes from the time I first heard him, he flies off and almost immediately disappears.

After the woodpecker's departure the silence is complete. Winter here is like that, a series of isolated, unpredictable events separated by greater or lesser intervals of time and space which seem utterly lifeless.

The trail winds over the undulating land, keeping for the most part to ridges, although now and then curving down to skirt the edges of ponds, low sedge meadows, and bogs. I find few fresh tracks, but a moose had preceded me down the trail. Farther on weasel tracks cross in front of me. They weave between shrubs, under half fallen trees; pause and circle where there must have been a warm scent. I can visualize the animal, black tail tip, eyes and nose dark jewels in a mantle of white, on its burning quest for another warm-blooded life to sustain its own. I almost expect the snow to have melted under its feet, but they have left only shallow dents.

I walk through silence for over an hour, then stop in sunlight to eat my lunch. The faint warmth on my face is a welcome balm.

Descending a short but steep hill, I push into a stand of white spruce through a prickling phalanx of saplings. Like hoary guardians the parent spruce spire high above them. At the heart of the stand, here on the north side of the hill, I am in another realm. No sunlight penetrates and the air is abruptly colder. The snow lies thinly and evenly as icing on a cake. Few shrubs tolerate the year-round shade and thick mat of needles that take decades to decay. The mature spruce are far apart; some have already fallen. One lies prostrate, its entwined, shallow roots seeming to claw the air in a futile effort to right itself again. Another has broken off a few feet above the ground, its dead heartwood ant-riddled. Some old balsam poplars remain, their hard, deeply grooved bark contrasting with spruces' thin, flaky texture. But they are headless columns, their crown branches hidden among the conifer foliage.

Since entering the spruce I have been aware of an almost continuous gnawing sound. I know what it is, but it is a few minutes before I locate the red squirrel sitting motionless on a branch. In the cold the squirrel is as compact as it can make itself—head

drawn in, tail clinging to the curve of its back, hind feet nearly covered by the fur of its belly. Only the forepaws are extended to hold a spruce cone. The squirrel begins at the bottom, clipping off the inedible scales one at a time to reach each seed. When it has finished, the scales lie scattered below. The squirrel drops the slender core of the cone, then continues to sit still. Suddenly it scampers up the trunk and out on a limb to the end where the chestnut coloured cones hang from the twig tips. The branch trembles under the assault, losing half its load of snow. Then the squirrel's head and shoulders appear out of the mass of needles; it nips off a cone, vanishes. Soon I hear gnawing again.

Climbing the hill beyond the spruce is like surfacing from a great depth, up again into the normal world of air, and sun, and warmth.

As I pass a sedge fen, something moves among the willows on the far side. In a few moments a bull elk emerges. Massive tined antlers sweep back past his shoulders as he raises his head to test the air. In the sun, and against the snow, his tawny body glows and reddish highlights gleam in the thick, dark hair of his neck and lower legs. He has heard my steps, but has neither seen nor scented me. His ears work backwards and forwards; then he moves, one leg at a time, slowly. He is uncertain, ready to flee. Gradually, however, he relaxes and walks more normally, but still carefully. A few more steps and he is gone, disintegrating into the snow-foliaged forest.

The afternoon light is yellowing; the air is suddenly colder. I must turn back. I go quickly, and on the way startle a ruffed grouse who hurtles off through the trees in apparently panic-stricken flight.

This is the fourth day of ice fog; of silence almost suffocating, as if the air were filled with white ash settling slowly from a volcanic eruption. The tide of daylight seeps in each morning and ebbs away each afternoon whitely. Animal activity is suppressed. A few new squirrel tracks mark the snow, but the animals lack their usual ebullience. When I see moose, elk, and bison they appear listless, less wary than at other times. Even the chickadees are subdued. At dusk there is no coyote song; and later, no horned owl's hoot.

Wind had long blown the snow from the trees and shrubs before the ice fog began, leaving them dark and clean-limbed again. But the fog does strange things. It does not settle out like ash. On contact with plants it initiates crystals that grow, coldly inanimate, throughout the fog's duration.

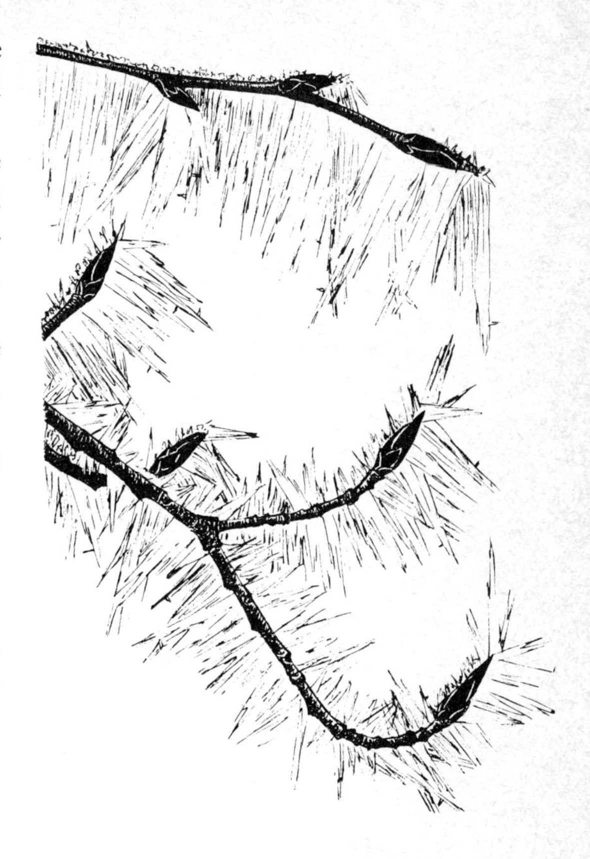

After the first day twigs, grass, sedge, and flower stems were white-rimed with thick, blunt crystals. Overnight from these, slender needles elongated. By morning every surface bristled. Around me massed a hostile thorn forest, but one whose defences were shattered by a careless brush of clothing.

There appeared to be no further change during the second day, but by the following morning the needles were longer and had proliferated. Slender twigs began to bend under their weight. Each scarlet rose hip drooped white-bearded. Spruce foliage hung matted as if with some quick-growing mould. The snow surface itself had become hoary, and on the rims of the wells left by animal footprints, knife-bladed crystals spoked inwards. Surprisingly, the tree trunks remained bare, even the corrugated bark of the balsam poplar. Only occasionally did the burl of an old wound support a meagre cluster of white quills.

In this morning's absolute calm the furring of ice has an architectural quality. The forest is again insubstantial, fragmented by white, but flat and dead without the interplay of sun and shadow. Towards mid-morning the chalky light brightens, while overhead faint smudges of blue appear. Gradually the blue deepens, spreads, clears. The fog thins away until it is no more. Snow and ice dazzle in the unfettered sunlight. I squint into the glare in spite of having been out for some time, and frequently turn to the dark blue sky overhead for relief. The sun is barely warm even when I face it, yet in half an hour it begins to loosen the crystals. They come away first from the tree crowns, in great chunks that fall slowly, crumbling to powder on the way. Through the afternoon the ice fall accelerates. By sunset most of the trees and shrubs are stripped. The winter-dark unity of the forest again walls the near horizon, while an inch or so of frothy frosting has been added to the ground snow.

❦ ❦ ❦ ❦ ❦ ❦ ❦

Drip . . . drip-ip . . . drip . . drip-drip . . . These sounds, gentle, soothing, greeted me as I stepped outside after breakfast. They came from snow melting on the roof, but there was little to see except the shadowy trees and the dusk of the snow. Beneath the sagging overcast the air, thick with moisture, slid over my skin in slow curls, but as yet no rain fell. Near zero the previous afternoon, the temperature had now risen to 35 above.

All morning I was restless; unable to concentrate on indoor work; eyes constantly veering to the window anticipating rain. By noon it was 40 degrees, and steadied there. The quiet snow packed underfoot.

The rain began. Hesitatingly for a while, then steadily, slanting before a breeze from the south. For a couple of hours it was disarmingly springlike. Water gathered and dripped from twig tips; glided down stems and trunks in dark streaks; haloed the scattering of grey willow catkins lured from their buds by October's last warmth. The air was almost scented. Then the temperature began to drop progressively. Yet the rain continued. As darkness came I could feel slippery ice forming on the shrubs, and sense physically more than hear the snow surface crack under my boots. At the same time my bare skin registered the rain's wetness. The boundary between liquid and solid was hairline.

Today's dawn vaults into a cloudless sky in a graded wash of colours that rapidly overcomes the night. During the darkness the rain turned to snow, and three inches of it now top the ice. As I step forward I am momentarily held; then the crust gives as my weight bears on the snowshoe. I brush away down to the ice. It is hard, its broken edge rough, but less than a quarter inch thick. Had all the rain come supercooled through air just below freezing, the ice layer could be half an inch thick, a file that would soon rub away the leg hair of elk, deer, moose, and bison.

Once again the forest is transmuted—now domed with glittering chandeliers of crystal, reflecting and refracting the sunlight, making it more confusing than dazzling. It is difficult to pick out the little band of chickadees moving at an angle to my own direction. They inspect the lee side of trunks, and the ice-free undersides of large branches. A blue jay calls in the distance, then comes closer, perhaps attracted by the chickadees' calling or my own crunching progress. Suddenly it is almost beside me, slipping a little as it swoops open-winged onto a poplar branch. Illuminated, the bird's varying blues are scintillating, its pale undersides almost opalescent. It gives one harsh *jeeeee*, then flies. It continues to call, distance mellowing the sound. Off to the right, farther away, a second jay responds.

I step over a fresh snowshoe hare trail, and a little beyond additional ones converge on a fallen poplar. Nearly all the crown twigs within reach have been stripped of their bark, and the bared wood gleams palely like the bones of an animal skeleton. The twig bark of downed live trees is relished for its high food value.

I continue for over a mile with no sounds but those I make, no moving life but my own. Even the air is no more than a coldness in my nostrils, as if my sense of smell had been anaesthetized.

I am almost past a clump of young white birch at the edge of an old bog before I become aware of a dark bundle halfway up the trunk of one. It looks like a squirrel's nest. Closer I make out the squat form of a porcupine. Its four legs embrace the trunk, but part of its weight is borne behind by a branch. With its head tucked down between its forelegs it seems to be asleep, and still makes no move even when I stand directly below. It has been in the tree at least overnight, for the snow around is unmarked. Possibly it has been there for days—nearly two feet of trunk bark have been deeply and completely removed. This will probably kill the birch, yet porcupines are not common and do much less damage to these trees than do moose who winter browse them, often breaking down saplings to reach the high twigs.

Coming out on the shore of the lake I pause. The sun against the cloudless sky appears motionless, but, turning my back on it, I watch its rapid course measured by the sundial boles of the aspen. A small rose bush is now in shade. Ten minutes later it is completely sunlit, its shrivelled fruits warmly red and shining. When I stare fixedly, I can almost detect the continuous movement that floats the shadows slowly to the right. The sun is the pivotal point of the day; the shadows extensions of its long, slender fingers of light; fingers that link us across almost a billion miles of void.

January

Each step is an effort in the soft, deep forest snow. As I pull up my buried snowshoe it rises mounded, heavy now after three hours of breaking trail. Too frequently hidden twigs spear the mesh, and I half stumble. But soon bright sky shines between the trees ahead instead of the massed dimness of endless trunks. Above, a gusty wind strides clattering through the crown branches. Yet here at ground level the air is almost calm, so effectively do the countless slender columns baffle the wind.

Abruptly, at the crest of a hill, I break out into the open. Below me the south slope catches the afternoon sun, its whiteness interrupted only by a scattering of shrubs. With unexpected force the wind plucks at my parka. But my face is protected by the hood's fur trim and I stand in a tent of warm air of my own creating. Comfortable for a time, I remain still.

The gust passes. Worked over by every wind, the snow is etched in the grained relief of weathered wood. The shadow lines are thin and hard and in places the bared ice crust glistens with a golden sheen.

I hear the next gust before it arrives. Then it swoops down over the trees and across the slope like a stooping falcon. The snow surface that appeared so stable is lifted in a streaming, obliterating haze. But, when the wind drops and the snow settles again, all is as before. The shadows, the curves are just as I first saw them. Another gust repeats the phenomenon; another; and another.

I find it almost hypnotic. Surely a gust will come after which I can detect some change, no matter how slight. I stay for perhaps twenty minutes, yet the streaming snow seems to alter nothing.

I shiver involuntarily, and the spell is broken. I am not dressed for prolonged inactivity—my toes and fingers are chilled, and the warmth around my body has cooled. I start for home, the falling sun at my back.

The year's life ebbs away. The feeble sun burns bright in the sky all its short day, yet even at noon cannot banish the edge of ice on the air.

It is two weeks now since I have heard coyotes sing; since horned owls have hooted. Chickadees are not to be found beyond the limits of thick, sheltering spruce. Silence has fallen over woodpeckers, blue jays, and magpies, and I rarely see them. Ruffed grouse exist only as explosions of feathers when I come too close to their snow dens. Small mammals—shrews, voles, and deermice—rarely travel on the snow surface. Weasels tunnel in pursuit of them. Red squirrels have retired in somnolence to well-insulated, mossy nests. It is now, far more than in the declining days of December, when some residual warmth from the departed summer still lingers, that I feel a twinge of fear. How rapidly doomed would be our earth should anything go awry with its orbital schedule, or the sun itself. No wonder sun worship was so prevalent among our long-ago ancestors. Indeed, I feel a need for it now; for some action that will ensure the return of spring, even though I can measure the daily minute increase in span between sunrise and sunset.

Late in the evening, under the twilight of the stars, I walk to the end of the wooded point that juts into the lake. The stars themselves stand close, their brilliant glittering and flaring threatening, at any moment, to scatter their light into the recesses of infinity. The whole universe seems on the verge of disintegration, and in a dead, cold world, I the only witness. Light glows in the north, strengthens, wavers like drapes being pulled across some celestial vault. For several minutes the unearthly, greenish-white radiance plays against the star-pointed night. Then, as quickly and inexplicably as it appeared, the aurora dims, vanishes.

The vice tightens another turn. Whiplashes whang across the ice as cracks split invisibly beneath the snow. Disembodied, the sounds arise somewhere in the distance, rush towards me, rip past, fade out. Inanimate forces are in ascendency, and for the present no living being challenges their power.

I am part of a landscape where life no longer seems tenable. During the hours that I trudge through forest I neither hear nor see active life. The snow compresses with thin, high protest under each step; while the branches overhead raise a death rattle as the wind tears through them. That is all. It is 25 degrees below zero. Tracks of many animals signature the snow, yet they give the impression of life gone, perhaps never to return; or even of clumsy, mechanical imitation to create the illusion of a populated land. All the tracks are blurred, are only identifiable from general shape, size, and spacing. For all I know the trees too could be artefacts, cast of metal or plastic cleverly textured and painted to resemble the real thing. It takes an act of faith to believe that their cells are only dormant; that buds exposed to all the abuses of this season can be called to life by April's warmth. Insects and other invertebrates crouch insensate, ameliorated in crevices, under bark, beneath leaf litter, down in the soil, and the mud of ponds; frogs, toads, and salamanders have burrowed underground; hibernating mammals are deeper still. Herbaceous plants have died back to ground level and are well-insulated by snow, but the vital parts of tall shrubs and trees must bear the brunt of winter's trials essentially naked—extremes of cold and wind, as well as the stress of rapid changes above and below freezing. Only their roots cling to the unchanging frozen soil.

In some ways the spruce seem more affected by winter than the leafless poplar and birch. They stand, branches interlocking, pinched with cold, an introspective look about them as if they had pulled their limbs closer to keep alive the tremulous flicker of their heartwood flame. The deep, vital green of summer is gone, and many needles have yellowed like those of an artificial Christmas tree that has been in the family for decades.

I come out of the trees where the shore is steep, and the wind slices across my bare face. I draw my parka hood closer so that its fur completely muffles my forehead, cheeks, and chin. But my nose still catches the air; the pain is real and sharp. I hold a mittened hand over it, prepared to endure the short-cut across nearly two miles of lake ice.

Segmented clouds span a third of the sky like flung-up waves of snow, wind compacted and sculpted. I watch for some time, but they do not appear to move or change shape. Like the real waves of snow on the ice they are frozen still, locked into the blue. Below the clouds the magnesium flare of the sun burns up the southern sky.

A mink trail meanders among the shoreline sedge hummocks and cattails. Beyond that stretches only the lifeless, wind-honed snow and ice. The repeated winds have driven the snow crystals so tightly into one another, transforming their delicate filigree to solid grains, that some alchemy seems to have taken place. I am walking, not on snow but pure white gypsum. My snowshoes leave no mark. I take them off and walk easily.

In the lee of a headland snow banks deeply and smoothly, the low-angled sun etching its stippled surface in blue and white. But near the edge of the lee, where the whirling snow was caught again and hurled onward, an earlier layer is exposed, rippled like the ropy surface of a soft lava flow.

Beyond the influences of the shore the pattern is more complex, and the whole history of the winter is recorded. Over the months a deeply carved frieze evolves. By tracing back through successive cross-grainings I can account for five major storms. In places the effects of the thaw and rain are revealed, where the snow is nubbled and glistening. Here and there snow has been swept clear and the ice is glassy smooth, riven by randomly directed cracks. Snow-filled, they resemble quartz veins through rock. Below the surface irregular conglomerations of air bubbles hang apparently suspended. Deeper still, the ice recedes to the same night as outer space.

Near the centre of the lake I cross a wild sea where snow waves curl in arabesques, and it is here that I come upon several poplar leaves trapped in the smother of white. Nothing in their brown, crumpled, distorted shapes links them with the living, and I am left to go my way more desolate for the momentary sense of kinship. Am I the last living being on this earth?

February

A week ago I crossed the last day of January off the calendar, but the sub-zero cold has not faltered. Twice during the past month the low has dipped to –52 degrees, fortunately in complete calm. At that temperature the air fills with ice crystals that sparkle against the sun, and remain suspended. Distant features blur in the haze; a suggestion of a rainbow haloes the sun; dawn and dusk skies are muted, the colours blending softly. This visual benignity belies air that sears as it is inhaled, and quickly burns bare skin to white numbness. No animal makes a move that is not essential for survival.

The cold has at last broken, and zero today feels almost mild. The tension eases; I dance, pirouette, leap with joy. A squirrel chittered this morning at sunrise; chickadees passed through the poplar around the house; jays and magpies called. But winter is far from retreat. The snow by now seems permanent, and it is difficult to believe it is maintained only by cold. It looks as if it has always been, could never melt.

Scarcely denting its solid surface, I ride the arch of a drift onto the lake. In the bay the snow is deep, but uneven in texture—here unyielding, there scarcely compacted. As did the mink whose course I parallel, I leave an intermittent trail, its blanks like gaps between the letters of Morse Code. Near the end of the point I pick up a coyote's track coming from land. At first the animal walks normally, but near a small island the strides shorten, then the footprints bunch. Three long steps, a space, a crater in the snow from a sudden pounce, a stain of red on white. This, the print of a small foot, and a diminutive, pale body feather caught on the edge of the hole complete the episode. The coyote's trail continues across the lake, while I strike towards another nearby island.

It is densely forested with mature white spruce, white birch and both poplars. In some places birch predominates, in others spruce, in yet others the poplars. The interior of the island is more or less flat; however, centuries of wave action together with changing water levels have shaped a steep shoreline, especially along the windward exposures. But there are a few notches from which the ground rises gently, and I make for the closest of these. At the marshy edge of the land snow billows over the massed sedge clumps. Its surface is pocked with a number of anonymous round holes about an inch in diameter. No tracks lead to or from them, for they are ventilation shafts dug from below by voles who live out the winter in the dusk and dark of snow-roofed tunnels. Occasionally, brown droppings or a slight urine stain reveal that a burrower has surfaced and perhaps paused to look over the rolling white desert. This is a venture the little rodent may never repeat—at night the movement, the scratch of claw on snow can bring instant death on soundless wings.

I move into the perennial chill of close-ranked spruce. The sun beats on the trees at the far edge of the stand, but only a few blades knife into its heart. Squirrel tracks criss-cross everywhere, usually in direct runs from tree to tree as squirrels are uneasy away from familiar escape routes. Around the bases of three patriarchal spruce, more squirrel trails overlie one another in trodden paths. Fresh cone scales litter the snow, and pave the numerous openings to a maze of tunnels. This area will be conspicuous even in summer. Generations of squirrels have used it for the same purpose, and now the residue of the cones is piled high. So frequently is new material added that no plants can gain roothold on the midden. As I inspect it the current owner, high above in the tented spruce, churrs disapproval of trespass.

I am surprised by the sudden appearance of four boreal chickadees. Their muted, husky lisping had sounded distant. Slightly smaller than the black-capped, they have brown heads, black throats, and dusky underparts. They are not common in the park, and I have never seen them away from spruce.

As I leave, a cascade of snow tumbles from a drooping branch and powders into a white haze. Soon another falls. Crystals softened by the sun-warmed needles are all the lubrication needed to release the trapped snow. The air today is calm. I look back. The thatched limbs return the light and mask from prying eyes the gloomy deeps within.

White birch soar high in graceful curves. Like seaweed they yield swaying to the wind. Perhaps because of this pliancy they show few scars until senility weakens them from within. The matte white bark and the ruddy twigs absorb the sun's essence and glow, so that looking at them I feel less cold.

Their inflected calls loud and positive, several redpolls materialize out of the blue to feed on the birch seed catkins borne high up on slender twigs. Presently a drizzle of fragments sifts onto the snow that already, under the trees, is peppered with brown seeds and scales. For a few minutes the birds are silent. Then one gives a querulous note; soon the whole flock joins in. Abruptly they are airborne and away, their fading calls burred and rolling like their flight.

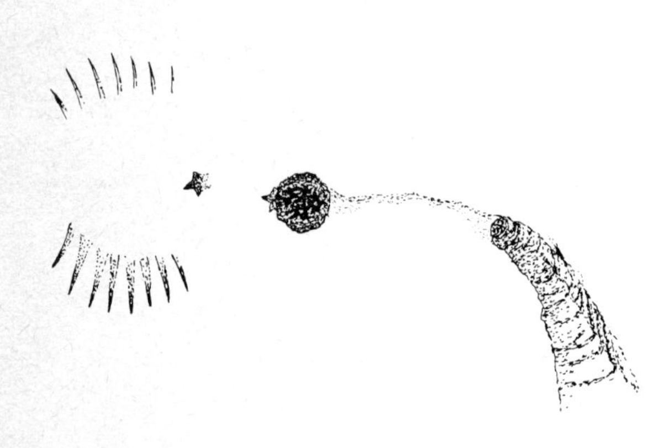

I cross the island. On the way a ruffed grouse rockets up a few feet ahead. As usual, I failed to see its head peering periscope-like above the exit of its snow den. A shallow trough becomes a tunnel where the grouse flew headlong into the snow, and I wonder if it could have avoided a hidden branch? After a couple of feet the tunnel curves sharply, and at its end a compartment the size of the bird's body is hollowed out. The bottom is layered with fibrous droppings. The exit is a small, round hole, and just beyond lie the lightly indented shadows of the feet and fanning wings.

A moose has visited the island often this winter. I see tracks both recent and old, and many browsed hazel and poplar saplings.

Leaving the island I circle around to my starting point on the lake. Near the site of the coyote kill I find the wandering trail of a deermouse. But instead of turning back to the shrubby shelter of the islet, or to the larger island, the trail veers towards the open where the shore is at least half a mile away. Deermice are forest animals, and while they will travel short distances to new home ranges, even to the extent of swimming narrow channels of water, they are more or less sedentary once adult. Winter particularly, curtails their activities. As soon as the snow is deep enough they prefer to tunnel through it, as its insulation enables them to survive severe cold. But perhaps the mouse did not go the entire distance? Or was caught by some predator? Several times I nearly lose the trail since the most recent snow has collected only in the lee of old, iron-hard ridges. But always I pick out the delicate tracery of the footprints again, following a remarkably direct line, almost as if the mouse had taken a bearing before setting out. Finally reaching land, the mouse had bounded over the snow until it entered dense shrubbery, then had promptly burrowed down. I am at a loss to explain this mid-winter odyssey. Why should the mouse have left the security of suitable surroundings on a venture that would seem, at the least, unpleasant? Had its food come to an end it would have had to travel only a hundred yards to the large island, or little farther to the nearest mainland. No other mice have made this journey, so there cannot be a general food shortage. The secret will remain forever with that single unconventional individual.

The sun has just trembled down behind the black fretwork of trees, leaving a burning gash on the horizon, when the first wailing note ascends. Still rising, it breaks into several high-pitched yips. A second, and perhaps a third, or more coyotes join in—each on a slightly different pitch and sequence. Communally the chorus becomes unordered, without pattern, yet never discordant. The wild voices are like the excited cries of children at play. Abruptly they stop. A minute or so later they flare again into the darkening sky. This time, as if in faint echo, others begin calling off near the south shore of the lake. Brief silence. Then the first group breaks out again, although now they are more distant. One coyote continues after the others stop, the yips faltering as if it realizes it has missed a cue. Then the silence stills, and grows deeper with time and my dwindling expectation of it being broken. The coyotes are off on the first hunt of the evening.

Here where deer, elk, and moose are abundant, coyotes probably form permanent packs like those of wolves, although they will often hunt alone for small prey such as mice and voles. Regardless of broader social organization, mated pairs are the basic unit and are unions which endure until the death of one partner. The chorus serves at least two purposes—to reunite separated individuals, and to advertise the group's territory. In all seasons it is concentrated at dusk and, to a lesser extent, at dawn.

As I turn eastwards I am momentarily startled by the great orange disk suspended behind the trees. Near the horizon, the moon is a painted circle on the sky. Soon, its power restored, it will trace its brilliant arc through the darkness as shadows of the sun's reflection dial its passage.

A great horned owl calls. The resonant *who-who-who-hoo . . hooo . . hoooo* rolls into the night. It comes a second time. Suddenly from close behind, a reply. The sound is very loud and vibrant; flows out and fills the emptiness. But when it stops, it might never have existed. The next hooting, minutes later, is from two birds both far away.

Early this afternoon I chance by a blowdown of aspen. Three mature trees have fallen, taking young ones with them, bending over and twisting saplings and shrubs. Close to the ground the snow has mantled the trunks and filled the hollows. Recent weasel trails web the whiteness, coming to foci at several tunnels which bore deeply into the piled debris.

A weasel's head and forelegs materialize at one of the holes. The animal is just suddenly a fact, its snow-white fur haloed by the dimness of the tunnel's interior. For several moments it is motionless, looking at me directly, calmly. Its composure is complete, expressing neither fear nor hostility, but rather an openly curious interest. The weasel tilts its head to examine me from a slightly different angle. Then it vanishes. Soon, like a jack-in-the-box, its foreparts pop out of another hole. It scrutinizes me again. A second time it pulls back abruptly, as if on an elastic band. This same manoeuver is repeated several times until the weasel's curiosity is exhausted.

I am compelled to reassess my impression of weasel character. The animal I have just met seems like any other predator, not the demon killer of its popular image. My earlier visualization of the little hunter "on its burning quest" is clearly a distortion.

The late February sun is genuinely warm. I stand facing it, slack-muscled, eyes closed against the glare. I am reassured that spring will come again.

For long minutes the usual silence suspends time and life. Then a woodpecker rattles loose curls of bark on a birch. I open my eyes to a female downy woodpecker, a plain black and white bird. She continues to batter, but the bark is resilient and her beak bounces off. She hitches up farther, makes a few exploratory taps, then begins drilling in earnest on a solid surface. Chips of amber wood flick out; the hole grows. She pauses and probes, withdraws her beak, and flies to another birch. I can not see if she found food, but probably she did. The tree is old, with several dead and broken branches.

I keep to a trail I have used several times during the winter and which is now well-packed. No new snow has fallen for over ten days, so the tracks of moose, elk, and deer are everywhere. Occasionally they show smears of blood; the ice crust, still there, has had its effect. Some of the animals have followed the convenience of my trail for short distances, the lighter deer and young elk not breaking through the pack. Where they have diverged I am led to willow thickets and the fresh white ends of browsed twigs—in places outnumbering the untouched ones. In the forest, sapling poplars in addition to hazel, buckbrush, saskatoon, and dogwood have been similarly used. Only the prickly rose and gooseberry twigs, as well as the less palatable bush cranberries, buffaloberries, and currants are largely ignored. Along the margins of ponds fallen aspen have been gnawed by elk; and on south slopes where the snow is shallow, elk likewise have pawed down to the ground uncovering dry leaves, grasses, and the evergreen basal rosettes of bishop's cap. All three have been eaten.

Off to the left an adult white-tailed deer, probably a doe, labours belly-deep in the snow, snatching at twig tips. Then, at a patch of rose, she delicately nips off the last fruits. Red squirrels also eat rose hips, although they usually discard the pulp and skin, preferring the numerous seeds. Ruffed grouse flutter up from the snow, and snowshoe hares stand on their hind legs to reach the rich food. Deermice can be added to the list of consumers, and chipmunks before they retire underground for winter. I remain motionless as the deer edges away unaware of me. There is no point in alarming her and having to watch her heave through the snow in panic. Most of her fat reserves will have been used by now and she will continue to lose weight until new green growth appears. Animals that become thin too early are doomed, and will die from malnutrition, often with full stomachs. Woody browse, even supplemented by rose hips, is a marginal diet.

The deer moves on out of sight. In none of the past three winters have I observed these white-tails gathering in "winter yards." At most I have seen three or four together, but even such groups apparently wander more or less at random, continually breaking new trails. Perhaps it is the generally shallow, dry snow, or the almost uniform availability of food, or both, which modify their characteristic response to a severe winter. And it is only in years of very deep snow that they suffer. Yarding can itself be hazardous, for if all the browse is eaten in the tracked area before the snow melts sufficiently, the weakened animals may be unable to break through to a new site. One winter a small herd of elk were trapped in this way and several died of starvation before the spring thaw.

A broad, deep, hoof-flattened trench angles across my snowshoe trail. It was here last time I came this way, and has been travelled since. Beneath it is probably a summer trail cut into the mineral soil. Bison are conservative, and plod the same routes repeatedly, usually in single file. Since the latest droppings are today's, I turn off. Twenty minutes later, through the thinning trees I make out a mass of dark shapes at the edge of a small pond. Again I keep my distance. It is early afternoon, and all but two of the bison are lying down, chewing cud and basking. While I wait I eat my lunch and gratefully absorb the sun. Despite the warmth the snow is dry and powdery. Its water content is so low when it falls, and the air so dry, that days of heat would be needed to turn it to slush. Before that happens much of the snow sublimates, passing directly from crystal to vapour.

Two magpies converse nearby in short, querulous notes. Then one flaps across the pond calling harshly. It is followed in a few seconds by the other. Almost overhead a grouse feeds on aspen buds. It is not at home high up in the crown, yet carefully inches out on twigs that look too slender to support it. As far as I can see it is preferentially selecting the larger, more nutritious flower buds.

Almost an hour passes. One of the standing bison walks a short distance, lowers her head, and begins swinging it from side to side to push the snow away from the sedge hay beneath. A calf rises, goes over to its mother, and begins to graze the food she has uncovered. One by one the others lurch to their feet, separate, and start to feed. Now and then a brief flurry of pushing or hooking marks older animals too near one another. But the calves press close to their mothers, snatching mouthfuls of sedge as soon as they are exposed. Altogether I count twelve bison in this clan—four adult cows each with last summer's calf, two yearlings, one two-year old heifer, and a bull who is probably two or three (bulls rarely breed before the age of five, and once fully mature associate with cows and calves only during the rut).

I return to my own trail. At the next opening, a large sedge fen sparsely dotted with willows, two massive bull bison are grazing. I stop, but they have seen me, stare uncertainly for several seconds, then walk away one behind the other. With distance, the angle, and willows half obscuring their bodies, the high, tawny shoulder humps appear to belong to one animal, and briefly I am looking at a Bactrian camel. Where the bulls were eating there are large craters in the snow; and it is obvious why sedges, particularly the awned sedge, are favoured in winter—the basal few inches of most leaves are still tinged with green. Furthermore, the heavily used upland meadows usually have little on them for winter pasture. Several snowed-in craters lie nearby. Bison do not linger at one sedge meadow until they have eaten it out, but move on after a day or two, gradually working around an extensive circuit.

All day I have been seeing fresh coyote tracks and occasionally, where the snow mounds over a stump or small shrub, yellow urine stains. It is now and through March that coyotes pair and mate. Even as I register these signs excited yipping breaks out not far away. It is repeated, ending in a wail. But no answer comes and, for now, no further calling.

I wind through a succession of ponds, nearly all occupied by beavers and separated by low, forested hills. The broad flatness of the ice exaggerates the relief of the land which rarely reaches as much as thirty feet. Here and there a line of tall spruce rings part of a shoreline, or flanks a narrow lane of low ground between opposing slopes, but poplar and willow predominate. I search each gathering of spruce before approaching, yet still fail to see half a dozen elk until they trot away from cover. It is almost impossible to detect a motionless animal, no matter how large, whose form is dissected by vegetation.

I stand over a file of unmistakable, large, round footprints that press lightly into the snow and suggest an animal ambling serenely through the forest. Although the long-legged, short-tailed lynx may weigh close to twenty pounds, so broad are its thickly furred feet that it travels as easily over soft snow as does the snowshoe

hare, scarcely a quarter its size and weight. However, I doubt this lynx will stay long in the park, for I have seen no sign of hares all day.

As I consider the lynx's presence, black-capped chickadees come near. Suddenly one whistles the three-note melody that is their only song, and to which I easily fit, "Spring's coming." Well, it is, and today is enough to make me believe it has already arrived.

The afternoon wanes, but the sun still warms. Where it strikes an exposed slope reflection is complete and the spotlight glare blinding. Dense-grey, hard-edged tree shadows lie across the snow as if etched into its surface.

All sounds coming out of a background of silence contain an element of surprise, for they cannot be anticipated. Thus it is when two great horned owls begin a duet. Now is the beginning of their reproductive season, and aroused emotions prompt daytime calling. During the minutes that I listen the birds do not seem to change position, and it is finally I who increase the distance from the source.

March

Dawn awakes heavy-lidded, reluctant, and the lowering overcast soon snaps shut over the gleam along the eastern horizon. The wind increases, bullying in irregular, swirling gusts. In the open, snow is picked up and carried along in a white smoke. Before long new snow falls, intricately patterned flakes wildly riding the roller-coaster air currents. Twenty minutes later I cannot see the far side of the lake; then all the islands fade out. The sphere of the visible continues to contract until objects no more than a hundred feet away are indistinct. By mid-day the snow is four inches deeper and drifts are beginning to cross the road. The prospect, not entirely

unpleasant, is of isolation by blizzard. I settle to an afternoon indoors. But, little more than an hour later when I glance out, the islands are there again. Soon it is all over but the wind. The clouds are breaking, blue showing between their blurred edges. Patches of weak sunlight sweep across the lake like searchlight beams. The cold front has passed.

The tracks of small mammals and birds are blotted out, those of larger, half filled-in. On their windward side all the trees are plastered with snow. In the willows fringing a sedge meadow I am confronted by a huge hornet nest. It is nearly sixteen inches from top to bottom, a tapering head whose chin just clears the snow. The likeness to a head is enhanced by the snow which folds and falls to one side like a nightcap. The integument of insect paper wraps round and round, winding sheets on a corpse. And such the nest is. Its construction and use span but a single summer.

Late, just before sunset, a file of white-blanketed bison trails across a meadow ahead. They look cold and miserable, but I know they are not. The superior insulation of their underfur, from which no body heat escapes, allows the snow to lie unmelted. As for expression, bison rarely look anything but bovinely impassive after passing through the brief skittishness of calfhood.

Now that longer, milder days have come, trips of several hours are both possible and pleasant.

I skirt a pond with a big, snow-mounded beaver lodge well out from shore. Not a stick of its construction is visible, but on the south side a maze of tracks leads to a well-used hole near the top of the dome. I find it belongs to a mink who has dug down to the roof, and there hollowed out a snug, sun-warmed den for itself. This is the first time I have found a mink's winter retreat in such a location. Usually the dens are along the shore at the base of a large tree or under a log. But wherever, such shelters are temporary, and used only as long as prey can be found nearby.

I snowshoe for some time without seeing anything but tracks.

Once, far off, blue jays and magpies call, and as I round a small bog boreal chickadees lisp within.

Snow shadows have their own perspective. When cast from objects near at hand they are intense, their edges precise; but as their maker recedes they blur out of focus and weaken.

I leave the trail briefly, pushing through thick willows that border a sedge fen. The resilient twigs spring back sharply and I notice that already their colour is brightening. But they have been heavily browsed again this winter and dead twigs are numerous in the stubby clusters that evidence years of intensive feeding. Suddenly I step onto trampled snow. On this altar lie a backbone, part of a pelvis, and a hoofed leg. The skull and other bones are gone—dragged away. But the size of those left, and the dark, matted hair, identify a moose. Too weak to winter through, it either died alone or was pulled down near the end by coyotes. The bones are now picked clean; no shred of muscle remains. For weeks this one death has been transmuted into many lives: the coyotes have shared with magpies, blue jays, chickadees, weasels, shrews, a red squirrel, even mice.

On the way home I come up to a porcupine rowing slowly through the snow. Despite its short legs, it sinks little more than belly deep. Behind it a furrow lengthens, ploughed by the hull of its squat body, punctuated by the oar marks of its feet.

Yesterday I brought back a red-winged blackbird's nest from a willow near the lake. It was difficult to remove it without breaking either willow or nest, as the sedge leaves that make up the outer bulk had been tightly woven around several branches. The bottom and sides are reinforced with firmly packed roots and peat, while the lining is of fine grass leaves. I left the nest overnight in a closed box in the warmth, and this morning one small spider is walking about. I carefully dismantle the whole nest, but find only two dead flies and a tiny beetle. Not all bird nests are thick and soft enough to be potential winter hibernaculae, and I have been collecting those I find that seem so. However, they have yielded little in the way of living insects and spiders, and not many more dead ones. Most nests are probably too small and exposed to maintain the stable environment so important for the winter survival of many dormant invertebrates.

It is difficult to write about winter in proportion to its length without being repetitious. Each year is different, and the season is a dramatic one, yet everything has been reduced to elemental essentials, all the frills shorn off. Storms pass through from time to time, bringing a few inches of snow or, occasionally, freezing rain. Subzero cold slumps down from the Arctic, remaining for days, or even weeks, without change. In January the temperature seems to reach absolute zero, but in reality only rarely drops lower than –45 degrees. Permanent snow may arrive in the last week of October, or as late as mid-December; may all have melted by early April, or not until the end of that month. Despite its duration, undrifted snow rarely exceeds two and a half feet in depth. Light as air itself when it falls, the snow slowly compacts to half or less of its original volume. Enduring, plants and cold-blooded animals lie dormant, their losses not assessable until spring. Most birds migrate, and few visitors replace them from farther north. Several mammals hibernate, are locked under ice, or are active under the snow. Those that adopt none of these evasions have one, and only one concern—survival. The summer harvest is all that there will be to sustain them through eight of the year's months. This is the drama, and one that often seems overly long, as when a playwright, carried away by his plot, draws out the suspense until it is almost unendurable.

But now the vernal equinox has passed. On it a pair of hairy woodpeckers began excavating a nest cavity in a live poplar. After a week's labour they have drilled a neat, round hole more than an inch into the trunk. (A few days more, and for some reason I never discovered, the incipient nest was abandoned.)

Today is cloudless and windless. In the afternoon heat I take off my mitts, throw back my parka hood, and open the zipper. The snow is soft and packs readily into balls which splatter against the long-suffering trees. My aim is not good, but the targets are endless. Besides, it is the spirit that counts. Winter is on its way out and I yearn to speed its departure.

Chickadees sing persistently and, I imagine, with new vigour.

Three magpies flap towards me from across the lake. Two land in the top of one poplar, the third in a tree nearby. They laugh lightheartedly, the flat, empty-husked winter *cha-cha-cha* full-bodied again. Abruptly one bird leads off, the other two quickly following. But soon it perches, allowing them to catch up. All three are chattering excitedly now. Then the leading magpie is off once more, the performance repeated several times before I finally lose track of their progress. It is impossible to identify sex in the trio, but their behavior suggests courtship. Magpies begin nesting early—old nests are renovated, and new ones begun well before the end of March.

My thoughts turn to great horned owls, and I visualize the females high up in poplar and spruce, incubating their two or three chalk-white eggs. They sit there for nearly four weeks, leaving only briefly, not at all during a storm. Often the female will be fed by her mate. Bringing his beak-held gift, perhaps a vole, he presents it gently—not an unusual act. Predators use their formidable armaments primarily for defence and the capture of prey rather than for conflict with their own kind.

Freezing by night and thawing by day the snow has become granular. Crisped by temperatures that occasionally still fall near zero, it is slushy again by mid-day. The sun's heat burns through it, radiates back from the ground, and erodes from below. I touch the surface with the toe of my boot, and it collapses into a hollow. Even in the forest a general subsidence is obvious, and the snow is receding from the warmed trunks and shrub stems which are encrusted with mosses so vividly green that they seem alight from within.

Out on the point I disturb a moose. It lurches to its feet, pauses to glance backwards at me, then crashes off through the undergrowth. Reaching shore it starts across the ice and is suddenly graceful, long-reaching strides rhythmic, body proportions in harmony with its actions. But now it is in the open I notice that its coat is ragged, and grey patches show on shoulder and hips where the hair has been rubbed off. Winter ticks, which periodically infest moose, are the cause of this abnormal hair loss. During the fall the young ticks climb shrubs and when an animal brushes past, quickly attach to the hair to spend the winter in heated security, gorging themselves. It is not known yet if the ticks transmit disease; however, excessive numbers seem to weaken the moose and may contribute to late winter deaths. At the least the parasites are irritating. Where a moose has been rubbing against a tree I find swatches of hair, blood flecks on the snow, and many squashed ticks.

Later in the afternoon I encounter more moose, all in thick willows. Two of a group of four adults are lying down; at a little distance a cow with last year's calf browses without pause. None of these moose have lost hair.

High, melodious whistling announces the passing of pine grosbeaks. Then the little flock suddenly sweeps into the top of a birch. The deep rose of the adult male fires brilliantly in the sun. He bends forward, testing the remnants of a catkin, and I can just detect the dry, papery rustle of seed and scale being separated. The others, three females in yellow-suffused grey plumage, and two sub-adult males, burnt orange on head, back and rump, preen carefully. The male whistles again, three times, the middle note lower. Abruptly they are all receding in flight, the first grosbeaks I have seen since late December.

A hairy woodpecker *pink*s nearby, a sound that hammers on the silence, although it is not particularly loud. The call is repeated several times, and I hear a second bird, possibly its mate, responding. Then the nearer one drums on dead wood with a rattling imperative. Hairy woodpeckers are year-round residents, and once established on territories usually indulge in a lengthy spring courtship.

At the top of a hill above a big beaver pond I come upon a red squirrel nest in a poplar, the only such one I have seen. The structure is bulky, nearly the size of a bushel basket. A sparse beard of dry grass hangs from the underside, but most of the nest seems to be moss. It must be occupied since fresh tracks lead to and from the tree, yet no spruce grow within sight. Here where there is choice permanent homes are rarely established in poplar forest with its often meagre winter food and shelter.

As day blends to darkness, and dimensions flatten out to two, then one, coyotes begin chorusing. It is one of those animal expressions that never fails to arouse the thrill of a first hearing. One solos. Low at first, the intonation rises and falls, rises and falls in a cadence that is never false, and so gradually dies away that only slowly do I realize it has stopped. Within seconds excited yelping erupts, the one now lost in the group.

April

The snow sinks down. Trees stand in brown wells, last year's leaves felted and heavy with water. When I put my ear to the snow I can hear the slow trickles of water drops filtering their way to the ground, a gentle wearing away of the past.

Patches of greenish slush appear on the lake ice, widening daily.

Lines of capped pedestals, several inches high, traverse ponds where moose, elk and bison crossed during the winter. The compressed snow takes longer to melt, and the thaw has turned the tracks inside out.

But not all movement is downwards. Tiny wingless snow fleas crawl from the leaf mould and smudge the surface of the snow with soot during the warmth of the day. As the afternoon cools, they withdraw. Other insects arouse themselves, mainly little flies. Some creep on the snow, but most float in the air like puppets on strings. Spiders, all small and dark, scuttle about, and once I see one pounce on a sluggish bug.

Sap is rising. The willows flush with a vividness they possess at no other time. Their flower buds are swollen to rotund fullness, and many catkins have already pushed aside the scales that capped them for over half a year. The thin, almost transparent aspen bark tints a brighter green; crown twigs curve, beaded with flower buds puffed to bursting.

During the day little flocks of pine grosbeaks and redpolls fly overhead without stopping; once fifty Bohemian waxwings. I hear the nasal *naa . . . naa . . . . . naa . . naa*, of a white-breasted nuthatch, but the bird flies before I can find it. Many black-capped chickadees have dispersed back into poplar forest. Some are singing. Reacting to the warmth and sun, hairy woodpeckers are courting actively. Their drumming tattoos begin with a flourish, then rapidly die away to nothing. Once two come close in flight, the female leading. She swoops up to momentary immobility on a bare trunk. The male lands likewise on an adjacent tree. Both birds become excited, uttering squeaky calls and hitching upwards with spasmodic, exaggerated jerks. Then the female flies, her hopeful mate close behind.

I flush a ruffed grouse from open hazel undergrowth, and it bores off through the trees like a missile. I should have seen it sooner; it certainly would have been aware of me for some time. But grouse always wait, to surprise and disconcert the intruder.

On a steep, south slope where the snow is patchy, three bull elk browse the low shrubs and pull at the flattened grass. Their movements are methodical, swaying heads sometimes synchronized in soporific rhythm. Only one still carries his antlers, although elk usually shed theirs by the end of March, two months later than moose and deer.

Walking is difficult in the soggy snow. It is too wet for snowshoes, yet still several inches deep in the forest. Along my own winter trails I slither off the uneven pack, and break through where I expect to be held. Recently travelled bison trails are the easiest routes to follow as the animals have pounded the slush firm again. But too often these lead where I do not wish to go.

Deep in an alder-grown depression a moose is bedded down.

It is barely a hundred feet away, but makes no effort to rise. Its head swings towards me, and its ears slowly cup forward. From the head it is hairless all the way over the shoulders and back. Its frame is more angular than it should be. I doubt that it will live to taste the first green growth its wasted body craves.

Dawn now is compressed, and the sun rises more quickly to warmth.

An hour or so after sunrise I watch a magpie flapping across a pond, a long twig dangling from its beak. For a time no other bird activity occurs nearby, so it is on silence that the first crow caws sound, throaty notes my ears have been tuned to hear for the past few days. I strain to see the birds, but they are too far away, or low in the trees. Later, in the early afternoon, two land in a poplar along the lakeshore. They *caw* briefly, then plane down to the ice and walk about picking here and there at things on the surface. Against so much white their black is intense, a negation of colour that is alive with sun-sheen. After a few minutes they lift off and beat towards the far shore, calling as they go.

I walk back from the lake and through rolling upland for over a mile. Before I reach the grassy opening I can hear small birds chipping. On bare, leafy ground a mixed flock of tree sparrows and juncoes is feeding, pecking among the leaves and occasionally flipping one over. The juncoes also scratch, jerking back with both legs simultaneously, too quickly for the eye to follow. Suddenly they startle, and flush, as one, up into the shrubs. The juncoes' fanned tails flash white along the edges, but the predominantly brown sparrows are hidden amongst them. All remain silent and unmoving for over a minute. A shrike or hawk must be close, but I cannot find it. First one, then another bird moves slightly, ruffles its feathers and preens. Then they begin to call again, and soon return to the ground. All but one, a tree sparrow who stays on his perch to sing a soft, dreamy warble. Perhaps stimulated by the sparrow, a junco flies up into an aspen and emits a dry, rapid trill.

As I turn to leave, a flicker of blue out in the meadow catches my eye. Phoenix-like, a mountain bluebird rises from the dead grass and alights in a sapling. He is a self-contained entity, his vivid plumage releasing him from the safe bondage of camouflage. Above, he is the blue of pure water reflecting a clear sky; below, the colour has run into the white feathers tinting them delicately.

To be in the forest now is to be entombed. High overhead the barren crowns of the poplar interlace, cutting the sky into leaded panes of blue glass. None of the mature trees is without blemish, and there is nothing to hide the flaws or divert my attention. Frost cracks gape; the spore brackets of white heart rot protrude, sometimes four or five on a single trunk; here and there bulge the blackened, rounded tumors of poplar canker. Where elk gnawed the soft bark years ago, the scars are dry, hard, and dark, for the original bark never regenerates. Lesions from countless lesser injuries mark the trunks. Some are top-broken, snapped off at a weakened point by a storm. Others, dead, incline against the living, upper branches entangled. Bereft of contouring leaves, shrubs expose spindly, winter-starved skeletons that often interlock almost impenetrably. Withered and bleached, the dead stems of asters, fireweeds, grasses and other tough herbs I can no longer identify lean on each other, against trees, or lie felled by the weight of snow. Fragments of peavine cling to shrub stems and swing in the breeze, their connection with the ground long severed. Where the snow has retreated quartets of dull crimson bunchberry leaves droop from dead stems. No new green spears through the matted leaves, layer upon layer deep. Patches of bishop's cap, wintergreen, strawberry, and twinflower lend merely the illusion of fresh growth. Plants require time to respond to the overtures of spring.

From the distance a fluttering approaches, veers away, descends, circles, and finally arrives at where I am standing. It is a white admiral butterfly just out of hibernation. Wings and body are clerically rich and dark in hue. Against this the broad white stole extending from leading to trailing edge of each wing is dazzling. The butterfly would seem to be an easy target for any predator, but perhaps its taste is repulsive. Regardless, its unpredictable, wavering flight will give it a certain advantage.

Before I reach home I see two more white admirals, and another overwintering adult butterfly, the mourning cloak. It too is dark, but with overtones of deep red. The marginal wing bands are pale yellow, and along their inner side curves a row of small blue spots. Although more colourful than the admiral, it is less flashy. The one I am watching alights on a twig, closes its wings, and disappears. In its stead a frost-blackened leaf stands tethered.

Uninterrupted sun for the past several days has reduced the snow to scraps lying in hollows and in deep shade. The surface leaf litter is saturated. Everywhere water gurgles downslope, flooding the sedge fens, forcing the voles to higher ground. All ice is dark with melting, and slits of water show around the edges of a few ponds.

The temporary streams flow full and swift. They erode the lake ice and where they enter, up to an acre of water has been opened. In Astotin Lake hordes of sticklebacks throng up the streams, reveling in the oxygen-rich flood after months of near deprivation. A scattering of long-legged water striders and shiny black whirlygig beetles skate and gyrate on the surface where the current eddies. Below, water boatmen, backswimmers and various kinds of water beetles weave back and forth, up and down, along paths that to me seem purposeless. Keeping closer to the bottom than most of the insects, pale, fresh-water shrimp scull in all directions.

Wherever I look now grey-woolly catkins are poking out from aspen buds. At a distance willow thickets seem white-frosted. The naked pendants of alder, stiff rods since their formation last summer, now dangle lax and elongated.

Eagerly I peer at other shrubs and trees, but their buds are still held tightly in dreamless sleep. In summer, leaves are the basic key to identification, supplemented by flowers and fruit. Now I must rely on buds, twig colour and appearance, and branching geometry. Even so, I have to lump the several willow and currant species, their subtle differences being difficult to catalogue. But I name the others easily.

Hazel twigs are chestnut coloured, the young ones covered with a fine pubescence; the plump, downy buds are slightly paler. In early May some of these will open, not to release furled leaves, but slender, red pistils which spread like sea anemone tentacles, awaiting the drift of golden pollen from diminutive male catkins resembling those of the related alder and birch.

Wild rose needs little description. Its spiky, bristly red canes, and tiny scarlet buds are badge enough.

The gooseberry's prickly, pale straw stems arch and sprawl weakly from a central point of eruption. Its off-white, pointed buds are covered with loose, papery scales.

The two bush cranberries can be separated partly by habitat, since the "high" prefers moist ground, and the "low," well-drained forest soils. Furthermore, as the names imply, the high grows taller—up to twelve feet—while the low rarely exceeds six. Nevertheless, there is much overlap. Both bear large, full, blunt-tipped, bright crimson buds arranged oppositely on light tan twigs which turn grey and old-looking after a couple of years. But the twigs are thicker on the high-bush cranberry, and where they branch are often conspicuously knuckled like arthritic finger joints.

Saskatoon, a tall shrub, is characterized by its shiny, reddish-brown, attenuated buds whose large inner scales are fringed with white hairs. The slender, reddish twigs, when over a year old, turn scurfy like peeling, sunburned skin.

I can instantly distinguish choke cherry from pin cherry despite their close relationship. The sizeable choke cherry buds are long-pointed, their pale-margined scales chestnut. Last year's twigs shine rich brown; all others have aged to flannel grey. In contrast, pin cherry buds are the merest knobs of matter, while the twigs retain their deep-red colouring for several years.

Buffaloberry clumps are untidy. Their pale greyish twigs; small, flattened, leaf-like buds; and grape clusters of pin-head flowers are all heavily stippled with minute, rusty specks.

The maroon-stemmed, wide-branching red-osier dogwoods mingle with willows, alders, and high-bush cranberries. Their bark is smooth and almost satiny in texture even on the main stems. The small, tapered buds are opposite, hoary with down, and no more than faintly red-blushed.

In contrast to the latent life of the forest, meadows seem condemned to eternal death until I part the shrivelled strands and discover the first green lancets of new grass.

The air is chill, on the edge of freezing. A northwest wind blows raw and gusty from before the first light of morning, harrying the squall clouds and speeding them downwind. On the horizon the clouds look close-knit and inconsequential. But as they near they swell, fray at the edges, reach over the forest, blot out the sun. Wet snow spatters down; thinly at first, then almost blindingly, driven fiercely to the ground. Soon the grass and leaves begin to whiten; my clothes are white-rimed; my face and hands run with melted snow. For perhaps twenty minutes I am cocooned in swirling snow. Then, as abruptly as it began, the squall is over. Space balloons out around me; the sky overhead turns blue; the sun reappears and dazzles. But not for long. To the north the white, brushed out edge of another cloud advances behind the wiry net of branches.

Throughout most of the day shower succeeds shower, although they gradually weaken, and by late afternoon the sky is almost empty. But the ground is all white again, while the old, pocked drifts have been brought back to winter. The temperature drops below freezing, and the descending sun fails to warm.

In the mild, sunny stillness after sunrise the now numerous crows caw hoarsely. From the end of the point I watch two in steady flight towards the north shore of the lake.

A loud drumroll sounds a fanfare for spring. After a half-minute interval it comes again; and repeatedly, the frequency remaining remarkably constant. This gives me time to locate the bird itself, near the top of a dying balsam poplar. One half of the split trunk, long dead, provides a fitting sounding-board for the big, flame-crested pileated woodpecker. He drums once more, and in the pause leans sideways, listening. I hear nothing. Perhaps he does, or perhaps not. He drums yet again, waits, listens; drums—for two minutes, then five. The last pause lengthens until I am drawn back to the tree. The woodpecker has gone.

The conversation of Canada geese comes steadily closer and lower. I look for the birds in the open sky, but find them only when they are pale against the massed spruce of an island, coming in to land on the ice. There are only four, although their unbroken calling gives the impression of many more. Once on the ice they pair off and separate, a goose from one pair insisting on a wide distance by advancing on the other with outstretched neck and yodeled calls that are loud to me nearly half a mile away. Then it quietens and returns to its mate. Soon each pair begins a mutual display. Stepping sedately in circles they face one another, necks arched high, beaks inclined slightly down. Occasionally one reaches out and almost touches its partner. The soft conversation between them is barely audible. They continue this way for some time, then become restless and call loudly again. Soon they rise from the ice, slowly gaining height and just clearing the trees at the northwest corner of the lake.

Before the geese are out of hearing, three pairs of mallards speed in low and circle the point, looking for open water. The hens quack loudly, but, although the drakes' beaks open and shut, I cannot hear their muted notes. Finding nothing but closed ice along the shore, the ducks veer away and search elsewhere.

I spend most of the day near lakes and ponds. During this time the first robins appear, big males with puffed, russet chests. They are assertive in voice too, frequently uttering a hard, *kuk . . kuk . . kuk*, but no song. A few fly out onto the ice and hop over it as if it were grass, picking up dead insects.

At a place where the thinning ice along the shore is slushy and broken a muskrat surfaces and clambers onto firm ice dragging

with it a section of cattail tuber. It hunches there in the sun, its sleeked, wet fur glistening so that only in shadow does it show colour. Holding the tuber in its front paws it gnaws quickly until almost nothing is left, then drops the pieces and immediately dives for more.

As I watch a crane fly weave uncertainly through the shrubs, a movement flicks on the edge of vision—the dark head of a mink popping up through a hole. It holds a large stickleback crossways in its jaws. With a single, lithe movement the mink is out on the ice, only the tip of its tail trailing in the water. It chews and swallows the fish, turns, dives.

I take a trail that leads to several beaver ponds. At the first, beavers are out. The ice near the lodge has already melted. A beaver swims to the feedbed, nips off a protruding twig, then hauls out onto the ice to gnaw the bark. Once finished it returns for a second, and again a third twig. Another, smaller beaver, almost invisible against the darkness of the lodge, sprawls on its belly soaking up the sun.

The second pond shows no life, but at the third, although I see no beavers, five mallards fly over. A pair peels off from the group, slants down in spirals, then splashes into the water by the lodge. The hen's quacking subsides, the ripples fade, and the two float serenely, preening lightly as if the action were an afterthought. The ducks are relaxing into drowsiness when a second pair of mallards descends, only to curve skyward, put off by the hen's hysterical quacking and the drake's wing-beating threat.

Flocks of tree sparrows and juncoes continue to drift north, feeding as they travel. Once I see a different junco. His hood is black, back and rump are washed with brown, and the sides are pinkish. This is a race no longer considered a separate species from the uniformly grey and white, and here more common, slate-coloured phase of the dark-eyed junco.

A crow flies into the top of a dead poplar on the far side of the pond. It sits motionless for a few moments, then bobs forward, holding its wings close, but fanning its tail. It bobs four times, each time enunciating a loud, full *caw*; then pauses. The next bobbing series, of three, is accompanied by soft, low, bell-like notes. They seem to form in the bird's throat, rise, then burst like bubbles in its open beak. A second crow comes close, and soon both birds fly away together. It is only briefly in April that I can experience what must be the crow's courtship song.

The sun is angled low as I make my way along the southern shore of the lake. Here more mallards dispute rights to the few small openings. One pair is on the water as a second planes in and lands on the ice about fifty feet away. The newcomers approach gradually, the male leading. The hen of the occupant pair bursts out quacking; then both swim forward, heave awkwardly onto the ice, and waddle hurriedly towards the intruders. The drake carries the threat further with lowered head, outstretched neck, and quickened pace. Soon he is running. Unsure of themselves, the arrivals take off, but are back in less than a minute and land again. This time they approach diffidently, with many pauses. The two hens quack at each other, and twice the drake walks forward and threatens as before. But he does not run, and when the second pair finally stop several feet from the water they are tolerated.

Near home I pass within thirty feet of another mallard pair, sleeping on the ice, beaks buried deeply in their back feathers. They may sense my presence, but in their exhaustion sleep on, eyes tightly shut.

Dusk deepens, darkness welling up from the ground through the forest. Across the still bright western sky a killdeer arcs in falcon flight, its penetrating cry belonging to the solitary and the self-contained.

❦ ❦ ❦ ❦ ❦ ❦ ❦

A narrow moat of water now rings the entire shore. It freezes over nightly, but thaws again before mid-day.

This morning the clear ice is thick, and the many large air bubbles trapped on its undersurface are magnets to insects who cannot take oxygen directly from the water. A stream of water boatmen constantly rises to and descends from every bubble. One at a time they float horizontally at the interface for a few moments; then legs jerk convulsively, and they plunge away. From time to time the much larger backswimmers and water beetles also come up for air, but hang head down with only the tip of the abdomen exposed. A caddisfly larva, its worm-like body encased in a tube of plant fragments, crawls along the underside of the ice. On reaching a rock, the insect transfers to its smooth surface one leg at a time, then continues its slow way until it eventually disappears around the far side. It has taken the cold-drugged creature, sluggish even in summer, at least six minutes to advance a few inches.

A song sparrow flies into a sapling and moves up from perch to perch until he is almost at the top. Sitting erect with tail drooping, he pours out a simple melody: an introduction of two clear, minor notes; followed by a choppy trill; closing with four reedy notes of which the penultimate rises steeply. He repeats the song so often through the next few minutes that I have it memorized, and can hear it playing on in my mind long after the bird has flown off.

Out over the lake gulls cry, but I pick them out easily only when their white shapes glide in front of forest. Without a telescope I cannot tell if there is more than one species; however, at least some of the calls belong to ring-billed gulls.

Early afternoon and the moat ice has melted. Large, dark-brown water beetles mass at the shore, clamber out, and stiffly in their unbending chitin armour, make their way up twig and grass stems.

Occasionally one loses hold before its goal is attained and tumbles down into the matted grass. Undaunted, it immediately renews the tortuous upward climb. Once at the summit the beetles snap their rigid forewings open, unfold the membranous hind wings, then launch into space. They drop a little before gaining speed, but soon lift over the trees on their spring dispersal flight. The larval stages are entirely aquatic, and the long-lived adults also spend most of their life in the water, feeding voraciously on other insects, shrimp, and newly hatched fish.

Farther along the shore more thousands of water boatmen than I can estimate crowd at the edge, scrambling over each other in a seething congregation, some in, some out of the water. I am not sure of the purpose of this gathering, although possibly it is a mating swarm, as only a small number of the insects take wing, and many of these soon land on the water again and dive under.

An adult beaver swims past muttering and groaning to itself. At a poplar felled before freeze-up and now lying partly submerged it stops and begins gnawing the bark. Autumn-cut trees, apparently abandoned, are now eagerly sought.

The wind falls to a breeze and the afternoon is warm. I sit near the crest of an open south slope above a narrow, winding valley—a series of breached kettles that have filled in to sedge fens.

The sere herbs and wiry shrubs are alive with invertebrates. Half a dozen different kinds of small wolf and jumping spiders shuttle over the crisp leaves. The jumping spiders also hunt along the highways of the shrub stems, creeping slowly or remaining still, waiting. A tiny immature orb spider dangles on an invisible thread. Red mites, bright scraps of vermilion velvet, inch over prostrate stems and the mountains of crumpled leaves. Flies, ranging from delicate, gauze-winged forms to gross, metallic-bodied blue-bottles, arrive and depart on their separate ways. A large, dark ground beetle rushes into view; pauses, waving its antennae; then rushes on. Hundreds of bronzed flea beetles, barely a quarter inch long, climb the stunted snowberry and cherry. When disturbed they leap up to a foot without opening their wings. I am close to the bare dome of an ant nest that humps out of the ground like some peculiar geological formation. The ants seethe over it in endless comings

and goings. Several huge mosquitoes investigate me, occasionally touching my skin, yet do not bite. They are one of the few species which overwinter as adults, and are rarely numerous.

At a sustained rustling behind I very slowly turn my head. A least chipmunk, a curve of black and white stripes, pokes through the leaves at the edge of the trees. It moves cautiously, frequently looking up from its quest for seeds and insects. When only a few feet away from the forest it turns back. Then it climbs a hazel, sniffs the bark, twists its head to see all around, finally descends to the ground again. There it grooms before returning to its search for food.

I close my eyes. An elusive fragrance floats up from the earth and I am stimulated to breathe more deeply. Grass blades chafe one another with a silken rustle. A dead leaf that has clung all winter lets go. I hear it strike the grass, then scrape as it falls closer to the ground. A large fly buzzes into flight. Mallards pass down the valley towards the lake. My ears feel more than hear the sudden compression of air under each downthrust of their wings. The hen is silent, and it is the drake's gentle voice, on the borderline of being, which identifies the ducks. In the willows below a song sparrow sings. Crows, magpies, and a blue jay call intermittently. Behind, to my right, a flicker exclaims, *keee'uw . . . keee'uw*. The jay flies into a nearby tree, announces its arrival, *jeeea . . jeeee . .*

jeeee; then leaves. A mosquito whines in my left ear. I shake my head, and it flies away out of audible range.

From high above the petulant, kitten-cry of Franklin's gulls sinks down. They are the first of the year, and I nearly open my eyes to refresh my memory of their glittering white forms accented by black head and wing tips. The chipmunk comes close again, but now its movements are staccato. It is bounding.

Whup . . whup . . whup . . up . up . upupupuprrrrrrr. Subdued at first, the sound amplifies as the beating wings reach their whirring climax. But the ruffed grouse does not tire, and drums at short intervals for several minutes. Then, after a long silence, the drumming comes again from farther up the valley. Distance has robbed it of its emotional pitch, and I find it lulling. The sun is warm on my skin and closed eyelids. All is harmony. I drowse. . . . .

An archaic scream stabs the Eden peace, searing my ears. But hawks hunt with stealth out of silence, and this red-tail is probably doing nothing more lethal than calling to its mate. Yet the scream is elemental in its simplicity, conveys barely-contained violence in its tone.

I open my eyes. The sudden glare is alien, almost frightening; the dark womb of auditory and tactile experience torn open. I lie naked, new-born, surrounded by a world whose borders are infinitely far away. I am no longer the centre, but merely a mote in the universe. Then, momentarily the strangeness fades and the dimensions of visual space become ingrained once more.

The hawk flies from its perch near the top of an old poplar. Its wings beat slowly, shallowly, and almost at once it begins to ascend a thermal. As it banks on a curve the sun shines luminously through its rufous tail. A second red-tail appears over the trees also circling. The birds scream to one another. Smoothly they soar up, up until they are so small I can see them only because I have followed them. Once I glance away I cannot find them again.

It is almost an hour to sunset, but already in shadow the afternoon's warmth has evaporated and the chill is wintry. A cock grouse drums; mallards, pintails, and widgeons splash down into the moat and compose themselves for night. In the tawny cattails edging a bay a red-winged blackbird whistles from his perch on a tattered seed head. In the poplars above, glowing in the sun, three grackles sit quietly.

The sun rests on the horizon, then slips behind it. From a flooded sedge fen comes a disembodied *cree-ei-eik*. Now, in the third week of April, chorus frogs are gathering to mate and lay eggs. The first frog is soon joined by several others.

A killdeer passes over swiftly, crying its aloneness. A few minutes later a winnowing snipe materializes out of the light above, louder and louder in its fluttering, headlong descent. Its wing and tail feathers are the most primitive of wind instruments, yet they thrum musically.

I pick my way carefully in the dimming light. A sudden rush of wings over the trees, and four green-winged teal zigzag to the water. So swiftly and steeply do they drop that I expect them to plunge beneath the surface. But after an initial splash they float buoyantly and calmly at rest.

The past few nights have been above freezing. A number of small ponds are completely ice-free. Around each lake the moat is wide enough for a breeze to ruffle it. The ice is in retreat. Cracks widen; open water appears within the fortress.

With the first rays of the sun a robin sings tentatively, but the phrases are short and lack the rich modulations which will develop later.

I watch mallards, pintails, widgeons, gadwalls, blue-winged and green-winged teal. The mallards are the most numerous, followed by widgeons. I find only one pair of gadwalls. Almost all these ducks arrive already paired, but the few unmated drakes are aggressive, now and then attempting to abduct a hen. Huffed scuffles thrash the water, and often climax in long, fast flights with one hen leading three or four drakes.

Close to the ice-edge two common loons swim slowly, dive briefly. They are elegant birds, streamlined in form, strikingly patterned in black and white. The surface of their compact, water-repellent plumage gives the impression of smooth upholstery rather than overlapping feathers. These loons may be migrants,

or the pair that nests yearly in the park.

A scattering of gulls, ring-billed together with a few California and Franklin's, stand about on the ice. They preen or just look about as if they had nothing better to do. From time to time one or more lift off and fly leisurely to some other part of the lake. When they settle again, far away, their shapes undulate in the layers of warm air wavering up from the ice.

Small flights of Canada, snow, and white-fronted geese wedge northwards. Their yodeling cries range from faint tremulos to ringing imperatives, depending on whether they are high or low, overhead or off toward the horizon. Some Canada geese come down to rest and feed, and a few will stay to nest on secluded lakes.

It is sunny and warm. Isolated cumulus clouds develop and slowly drift downwind. I go for a long walk lasting nearly all day.

New sedge leaves tip above the flooded fens. From among these, camouflaged by the collage of bleached sedge clumps and burnished willow tangles all faithfully replicated in the water's mirror, a pair or two of mallards invariably start up before I see them—a volley of quacking and flailing wings. Frogs silence instantly in alarm, but soon those on the far side resume their singing. The shrilling chorus frogs have now been joined by the counterpoint clacking wood frogs.

I peer into the shallow, transparent water. Tiny beetles swim to and fro. A fat, cylindrical crane fly larva, resembling a piece of sedge stem, floats at the surface. The tip of its posterior, circleted with fleshy tubercles and stiff hairs, is exposed as stale air is exchanged for fresh. Then the cylinder convulses and descends, taking with it a shining bubble of air secured by the water-repelling hairs. As my perception of the small sharpens, I notice hundreds of tiny, black mosquito larvae. Most hang head down at the surface, but as I stir the icy water with a finger, those to feel the current wriggle away. They will develop slowly for some time yet, not emerging as winged adults until the middle of May. Far more minute than the mosquitoes are a variety of crustaceans which appear only as twitching specks. Water mites, several times larger, are yet little more than pinheads. Their eight slender legs would seem useless for swimming, but stiff hairs that are invisible without a microscope convert them to oars. The mites glide smoothly and rapidly.

Mourning cloaks, Milbert's tortoiseshells, and white admirals are flying by late morning. A Compton tortoiseshell, another overwintering butterfly, is one of a million leaves on the ground until its wings spread just before flight. Then the broken pattern of orange, brown, near black, and buffy yellow revives the vivid hues of autumn.

All the aspen are thickly tasseled with limp, grey catkins, edged in glistening pearl when I look at them backlighted by the sun. Both male and female are similarly woolly. The crowded willow catkins, like aspen, are grey wooled, but remain short and stout. The anthers, red-topped at first, elongate through the protective hairs before maturing. On the female catkins the pistils curve out, slender and pale apple-green. Alders flower less profusely, but their smooth catkins are now mature, the protective plates have separated, and the wind drifts the pollen to receptive female flowers on inconspicuous, crimson spikes. Later to bloom, balsam poplar buds are only now beginning to swell noticeably. Birch catkins remain unchanged, as short and stiff as when they were formed late last summer.

I sit at the top of a shrub slope eating lunch. A satyr butterfly pumps a dilettante course just above the shrubs. Its blended amber, orange and rust are showy in the sun against its drab surroundings, but must have been camouflage as it sought a hibernating retreat. With its irregular wing margins it appears battle-scarred; however, these are natural and enhance the dead leaf disguise when the butterfly rests. Soon after the satyr passes, a ladybird beetle alights on a nearby twig. It tucks in its membraneous hind wings, and closes the domed, scarlet-enameled elytra over them as we would shut a suitcase. Four symmetrical black spots are the only adornment. The beetle crawls steadily up the twig. It will be at least a week before any aphids appear, and until then the ladybird will fast, or nibble on pollen.

Suddenly a doe white-tailed deer steps out of the trees. She is thin, and her coat ragged with shedding, yet she retains the unconscious grace of her species. An eddy in the breeze carries my scent to her. Instantly she whirls in my direction and snorts—a coarse, trumpet-blast alarm. But she looks at me without seeing. I am half hidden and motionless, and she cannot separate my form from its background. Scent is enough. The doe stamps a front foot, snorts again, pivots, and bounds away, tail raised in a white flag of warning. The slope seems emptier than before her coming.

On nearly level ground at the edge of a poplar grove I look down on several mounds of fine, loose earth, each about a foot and a half in diameter and several inches high. They are not ant nests, but the work of pocket gophers. Unlike most other burrowing mammals who use their underground dens mainly for refuge, the gopher, like the mole, is truly subterranean, and surfaces infrequently, usually under the cover of darkness or snow. It excavates a maze of shallow tunnels to find the plant roots, bulbs, and tubers which it eats, then periodically pushes out the surplus earth onto the surface. As the exit is then almost invariably plugged, no obvious clue is left to suggest the origin of the mound. These little rodents do not hibernate. In some parts of their range they build nests under the snow, and deposit ropes of dug soil along narrow snow tunnels; however, I have found neither in the park. Here where the ground freezes hard, the gophers dig deeply in autumn and store a supply of winter food.

Every pooling of water is announced long before I reach it by the high-pitched throngs of chorus frogs, accompanied by a lesser number of throaty wood frogs. From a distance the sound seems to well up from the earth itself in soaring columns that fountain into space. Beside the water the singing becomes noise so loud and minor it brings me to the threshold of pain.

Along the mud-beached shore of a small lake several irridescent grackles strut, now into the water, now out of it, and probe for food. A flock of small, black birds approaches and flies into the poplars lining a sedge-filled bay. For a minute or so they chatter in short notes interspersed with squeaky attempts at song. Then, singly and in twos, they glide down to the half-submerged tussocks. Here they feed, balancing on stems and wading up to the limit of their legs. At last I can see them properly—the males black and glossy, but only faintly irridescent; the females slate-grey and dull. Both sexes, however, have the almost white eyes diagnostic of the rusty blackbird. Later today, having fed and rested, they will probably continue northward. Although the park is within their breeding range and seems to have suitable habitat, I have not seen these blackbirds here except during migration.

Farther along the lake in tattered cattails more than twenty male red-wings whistle and call, flutter from clawhold to clawhold, and occasionally chase one another briefly. The unseasonal heat stimulates the first singing. The birds' liquid *conk-a-reeeee* is the most musical utterance of any of the smaller blackbirds.

I turn away from the lake and almost at once flush a robin from a poplar sapling. As the bird flees protesting, I glimpse tawny underparts against pale fuscous—the first female of the year. On the way home I see several more. They are slender birds, and their movements already seem as furtive as if they were disguising the routes to their nests.

In the forest the drying leaves rustle under my feet, crisp and brittle. All surfaces are dessicated—lichens have paled and curled, mosses withered dry. But when I put my hand on an aspen trunk it feels cool and almost damp; and when I press the ground I find wetness quickly under the skin of dehydrated leaves.

Scattered cumulus, each one crowning an invisible thermal, texture the sky. Several hawks soar against the clouds during the afternoon. At least five are red-tails, but the others are too distant to identify. Two small V's of Canada geese pass over, and flights of ducks, ranging from a few to about thirty birds, are common. I inadvertantly startle others from their feeding in shallow water. Sail-white in the sun, ring-billed gulls beat slowly, graciously, to wherever it is they are going. At frequent intervals I hear crows, blue jays, and magpies.

After a long search I locate a hairy woodpecker excavating a nest in a large aspen. The steady tapping has a ventriloquial effect, and I examine many trees before finding the one with the bird. A little later I come up to a pair of chickadees investigating an old cavity in a rotting birch stub. Both pop in and out of the hole several times, then leave suddenly, breaking their silence with quick, conversational notes. While observing the chickadees I have been aware of a pileated woodpecker calling a long way off. The repeated *wick-u* syllables are each distinct, and without the flicker's rising inflection at the end of the series. Wherever there are willows and alders I see and hear song sparrows. At the edge of a grassy opening half a dozen tree sparrows scratch and peck in the leaves. They must be nearly the last of this spring. Throughout the day I am never out of the sight and sound of birds.

Mammals are more elusive, except for red squirrels. Once I cross a recently used bison track—fresh dung, deep hoofprints in muddied water, and a few tufts of wool underfur combed out by the rose thorns. But the animals themselves have gone on, and I am half grateful. Alone, I never relish a face-to-face encounter, for with no barrier between us these hulking bovines loom larger than life. They often seem docile, even indifferent, yet I always feel an aura of threat at close range, possibly because I am unable to read any mood in their impassive eyes and stance.

The late afternoon shadows are long and cool, but where the light still strikes even the ice looks warm. The humid air diffuses a golden glow. I near the lake and home. A grouse drums away in the forest; robins and song sparrows sing more strongly as the day wanes; a snipe winnows fervently, again, and again, and again.

At the far edge of the moat six mergansers swim and dive. One of the three pairs are common mergansers. The drake's impeccable black and white plumage is almost prim; even his mate's subdued chestnut, grey and white are neat and cleanly defined. In contrast, the four red-breasted mergansers are decidedly raffish. Both sexes have raggedly crested heads, the drake's being green-glossed black, the hen's dull chestnut. The drake's back and shoulders are dead black, his white underparts interrupted by a dark-spotted tan vest, and grey-vermiculated flanks. The hen is toned in blending shades of soft brownish-grey. Both species are migrants, so I watch them closely for some time. They appear settled and may stay overnight. When not diving for sticklebacks, they paddle slowly or occasionally stop to preen quickly. Finally I must move. The mergansers begin to swim away, ripples fanning out from their breasts, but they do not fly and I walk, I hope, almost imperceptibly. I reach the house without putting them up.

About to open the door, I glance across the bay. Ponderously heavy-winged, funereal, a great blue heron flaps southward over the ice, over the forest, beyond sight.

The temperature dropped steadily all yesterday, and thick, heavy clouds pressed down. I could feel their oppressive weight as the masses piled higher and higher. The colour drained out of the landscape, the glowing willows and dogwoods, the verdant young sedge. Most birds seemed to disappear as the wind came; they fed quietly or sought shelter. By mid-afternoon snow whirled and eddied, delineating the air currents. Soon it became a streaming river tossing and breaking over obstructions in its bed. As I walked home in the early dusk I was buffeted, and listened to the storm roar through the trees like a waterfall. Only the ducks, dark blobs turning, dipping, tipping-up on the steel-grey water were unperturbed.

During the night I woke more than once, each time to the ceaseless thrashing of the wind.

This morning comes dully overcast, the burdened clouds still low.

But the snow has stopped falling and now lies several inches deep on the ground. The wind has plastered it against the tree trunks and some has been caught in the baskets of the branches.

In the shallow bay behind the house a dozen snipe huddle on a ridge of bare mud. They all face in the same direction, although the wind has died. And they all adopt the same pose—heads pulled in so that they appear neckless; long, down-pointing beaks resting on their breast feathers. While I watch they make no attempt to feed, yet several red-wings and grackles walk among them probing the mud deeply. Mallards, widgeons, pintails, and teal, together with a few gadwalls, shovelers, and the first coots crowd the open water intent on finding food. A long way off two pairs of red-breasted mergansers, possibly the ones I saw three days ago, swim and dive. Near them several common goldeneyes, lesser scaups, and buffleheads dot the water. On the ice itself California, ring-billed, and Franklin's gulls cluster. They stand; wander about, occasionally peck at the ice; preen; flap their wings. They seem undecided about any further action.

On land, robins fluffed out to twice their normal bulk flock to shrubs and devour the tender buds. In a sudden reversal to winter behaviour, males in close proximity are amicable. A few days ago I observed the first territorial disputes. Song sparrows call from low down in bushes, but they do not sing. Once I put up two tree sparrows from the snow where they were picking at old grass seedheads.

The temperature rises slowly through the morning and by noon nears 42 degrees. The wet snow turns to slush against the ground. Lubricated by melting, it slips from tree trunks and, like soft-whipped egg whites, droops in festoons from the branches. The sedge meadows are opaque with waterlogged snow. The frogs are silent. Here where the transition from land to water is often obscure, this spring snow clearly defines the boundaries, being either immaculate white or grey-green.

At the edge of a bog I startle a snowshoe hare. Its mottled brown and white form, more brown than white, zigzags away through the growth, and is out of sight almost before I recognize it. This seems to break the spell of silence in the wake of the storm. A blue jay calls in strident notes. Within a minute I am upon a pair of black-capped chickadees gleaning an alder thicket for dormant insects. They show their usual élan, and chatter in a tumble of sibilants, gurgles, and hoarse *dee-dee-dee*'s. In their presence I almost forget the snow and the lowering sky.

Along the edge of an aspen grove a skunk's trail winds through the outlying saplings and shrubs. The irregular steps, footprints often overlapping, reflect the short-legged animal's ambling gait. Here and there it stopped to push its nose into the snow, and occasionally to dig to the ground.

The day ends as it began, under thick overcast with no wind, and in silence. As long as I look at one object I can believe I still see every detail; but once I glance away, then back, it has been reduced to a single plane. Finally even the clouds are only a featureless, ever darkening substance.

During the past two days the overcast has thinned and broken; the sun has eaten away at the snow. Much of the ground is bare once more and wet as after rain. Beetles, water boatmen, crustaceans, and mites glide and jerk through the cleared water. Hordes of mosquito larvae hang suspended from the surface film.

Early bird migrants are hardy, and I have found no casualties, nor seen any who appear weakened. This morning I watched a mountain bluebird hopping about on the ground and picking up torpid insects, but it ate them only after flying to an elevated perch. Under normal conditions these birds either catch insects on the wing, or drop onto them from hovering like a kestrel.

May

Warm air flows in on a southwest breeze. The sky is cloudless. The last of the snow melts and trickles away, or pools on the sodden ground.

The ice on the lakes is barely afloat, lead-grey even in sunlight. The many channels and ponds of water are blue like the rivers and lakes on a map, but the continent holds together only because no wind has ruptured its weakening bonds.

A flock of small birds sweeps in low across the bay and drops onto the ice. For a few seconds the birds cluster, then fan out—running, pausing; running, pausing. Frequently they peck the ice. Binoculars bring them close, and I make out the streaked, buffy underparts, white throat, dark greyish upperparts, and white-bordered black tail of water pipits. They are birds of arctic and alpine tundras in summer, and are now on their way to those destinations. Within ten minutes they all take wing again, punctuating their undulating flight with low, clipped *pip-it*'s.

Out of a tree behind pours a sudden, hurried warbling, and the next repetition of the song is just as hasty. I turn. Tasseled aspen branches frame a purple finch, the deep rose of his head and breast coming to life in the sunlight. As I drink in the heady wine of the finch's plumage and song, I register the asthmatic wheezing of a yellow-bellied sapsucker. The bird is near, but I fail to see it until it flies. From behind, its black and white markings resemble those of the downy woodpecker, with the addition of white shoulder-flashes.

Activity gradually returns to what it had been in the sunny, warm days before the snow. By early afternoon the frog chorus resumes, somewhat thinly at first, full-throated by sunset. As I pass a sedge meadow the combined voices of the little amphibians almost smother a loudly quacking dispute between two hen mallards. The variety of birds is steadily growing, but numbers remain small and I can still isolate the individual contributors to song and call.

The sun sinks into a crimson sea. Soon it is too dark to distinguish form below the skyline. From the lake a grating, braying rattle erupts in a shattering volcano of sound. Almost at once it is echoed by a reply. No other bird utterance can match that of the red-necked grebe. It gives no pleasure to the human ear, yet involuntarily you stop and listen. Perhaps in its timbre some ancient message vibrates—an antediluvian voice from before the evolution of song, an audible link with our biological past. It seems appropriate that I did not see the grebes fly in. It is as though they have been reincarnated, floating up through the buried millennia in the darkness below the lake to burst upon the present.

The open ponds lie as blue, grey, or sparkling eyes in the moraine depending on the day. But the lakes are still lidded and repressed under ice.

Today began cloudless. Now, early in the afternoon, dark-bellied squalls sail eastward on a constant course, rain drenching opaquely from their undersides.

Overhead the sky darkens, and the sun is quenched. Sudden gusts, most of their force downwards, thrash the branches. A few big rain drops splash down, spattering the dry leaves almost like hail. In a few moments the released rain drums. Streaks on the trunks reach the earth, widen and merge until the windward side of the trees darkens and glistens. Drops quickly form on twigs, fall, and are replaced in seconds. Soon the dead leaves are limp and yielding, absorbing the sound of the rain.

I turn my face upwards, hold out my hands from my poncho. For seven long months, since last October, we have been deprived of real rain. My skin is eager for its wetness, my ears for its liquid music that will not, in a few hours, be frozen to silence.

The wind billows like the clouds heeling before it. The lake stirs. The ice shifts. Soon waves are cresting white in the open and sliding under the ice in rhythmic surges. In an hour or so it is all over. The cracked, foundering shell fragments. Chunks pile up whitely on downwind shores. They are surprisingly thick, still several inches, but so riddled with holes that water pours through them almost unhindered.

By late afternoon most of the squalls have been driven away, and in the slanting light the lake is deep ultramarine silk wrinkled by the wind. The ducks and gulls have spread out so they are nowhere numerous. At last I can turn my back on winter. It is the second of May.

Two days after break-up remnants of ice still litter the shores, glittering in the sun like glass; but they will soon drip away to nothing.

Sunrise is so early now that I often luxuriate in bed for half an hour before I need to rise. Outside sounds come freely through the open window. Dominant over all are the red-necked grebes' stentorian outbursts. Two or three pairs usually nest in the bay behind the house, so interchanges are not only between partners, but also between pairs questioning territorial boundaries. As if in echo, during lulls among the near grebes others are clearly audible up to half a mile away. Nevertheless, when a robin carols from a perch beside the window I can hear nothing else. So near at hand and overwhelming, the song is almost unpleasant with reverberations masking its musicality, and I wonder how it sounds in the robin's own ears. Then the robin moves to a more comfortable distance and becomes part of the background medley which includes song sparrows, red-winged blackbirds, grackles, magpies, crows, snipe, ducks, a blue jay, a purple finch, and a distant flicker. Gulls call overhead as they circle, shrill, thin cries that impinge sharply on my ears. Once a flock of white-fronted geese passes over, their high, wavering notes seeming to embody the urgency of their northward flight to the Arctic tundras. Just as I am about to get up, a phoebe begins the slurred repetition of its name that serves it for a song. This is the first bird of the year, and it has unerringly found a man-made dwelling. Only the park's buildings provide suitable nesting sites; without them the birds would not come.

Outside after breakfast I find many more ducks on the lake than yesterday, but most well out from shore. Lesser scaups and common goldeneyes greatly outnumber all other species combined: redheads, canvasbacks, buffleheads, and ring-necks. The majority of the mallards, widgeons, pintails, gadwalls, and teal seem to have dispersed to the ponds. Those who remain have gathered in shallow weedy bays where nesting cover and food are close together.

Towards the centre of the lake a tawny sliver lies on the dark water, an island of giant reed grass anchored to the crest of a knob just below the surface. By mid-summer the flowering stems, topped by long plumed heads, will have soared several feet into the air. The stems themselves are stout and cane-like, and give off many broad-bladed, pointed leaves. A few grebes and coots, and a colony of red-winged blackbirds nest in the heart of the island, protected from storms and waves by the pallisaded growth surrounding them.

I hear a familiar fluid twittering and turning from the telescope quickly locate the source, half a dozen tree swallows conjured out of the empty blue sky. They seem more buoyant than the air itself as they swoop, twist, glide, and bank low over the water, moving closer to shore in their pursuit of small insects. One leaves the flock and perches in an apsen. It is as if a film had been stopped suddenly. The emphasis shifts from the swallow as an acrobat to the swallow as a bird. The blurred pattern of his plumage clarifies to irridescent, metallic blue on back and head, dull sooty on tail and slender, scimitar wings, and pure white underneath from chin to tail. For swallows, perching has become almost foreign and they have short legs with small, weak feet. They feed in flight and their courtship, though not actual mating, is also aerial. My swallow returns to his element and the film whirrs on.

As I look over the ducks once more a large bird flaps slowly westward just above the distant trees. It is too far for binoculars, and it takes me some time to pick it up in the telescope. I am fearful of missing it altogether, but just when I despair there it is, unquestionably an osprey. Perhaps when pike and suckers regularly migrated up Astotin Creek to spawn and summer in the lake, ospreys were regular visitors or even nested. However, dams and siltation have choked off the fish immigrants leaving only the very small sticklebacks and fathead minnows, so now this striking, fish-eating hawk is a rare transient.

In the wake of the rains the close-cropped meadow grasses are greening and the bison have found the new growth—many blades are truncated, the cut edges margined with the brown of dead cells.

A tide of yellow pollen is beginning to rise over the willows crowning the sedge meadows. Underneath, yellow-green and blue-green sedge leaves spike several inches above the bent and broken aftermath. Near the margins coltsfoot and marsh marigold have pushed up tightly-fisted flower buds.

Here and there, no more than that, bud scales are parting from the swell of leaves within. In the heat of the afternoon balsam poplar buds are sticky with tangy-scented resin.

As I enter a meadow several bison, mostly cows and yearlings, amble away from the far side. Bits of dry grass and smears of mud cling to their ragged, woolly hides.

After the bison leave I notice a male kestrel perched near the top of a dead aspen. He is the most vividly hued of the North American falcons, the initial impression being red, blue, and white. A circular crown patch, the back, rump an tail burn rich terra cotta. The rest of the crown, shoulders, and secondary wing feathers cool to smoke blue. A stain of buff crosses the chest, but the sides of the head, chin, flanks, and under-tail coverts are clean white. Black is an accent colour—a bar down through the eye, repeated in a second bar over the ear patch; two large, black ovals on a buffy ground on the nape which, viewed from behind, are like "super eyes"; the primary wing feathers; a broad band adjacent to the narrowly white-tipped tail; and spots aligned along the flanks. Finally, black spotting and barring crosses the shoulders and back. The kestrel permits a close approach, his attention being focused on the ground below. Searching for movement he swivels his head from side to side almost as freely as an owl. Three times he flies out over the grass and hovers for several seconds, yet makes no attempt at a strike and returns to the dead aspen. I try to edge closer for a long-lens photo, but this seems to unsettle him and just as I focus he flies, although not far. Again I gradually close in, and again he is off at the crucial moment. Twice more I attempt, without success. However, the falcon does not want to leave the meadow and is gradually leading me around its perimeter. I give up the idea of photography and leave the bird in peace. Memory on the film of my mind will have to suffice.

Later in the forest six cow elk startle away. Were it not for their light rump patches I might have missed them entirely, so quickly do they dissolve into the trees.

Contracted into the span of two days, sandhill cranes surge northward in wave after curving wave. No more than specks against the cloudless blue, too high to see as cranes even with binoculars, their resonant bugling drifts down thousands of feet of space, millions of years of time. Today the flights began in mid-morning and lasted until late afternoon. I heard more than I could see to count, but estimated at least two thousand birds. Once, shortly after mid-day, I looked up to find two great flocks breaking their steady course. For over a minute they eddied in smaller groups—a slow ballet of foam flecks alternately shaded and sunlit, pearl grey and flake white.

Together with an assistant, I spend the day on a waterbird survey.

The big duck flocks on the lakes are no longer there—have they continued on migration, or dispersed to the many permanent ponds? It is a question we cannot answer with certainty since ducks have been on the ponds ever since water appeared. However, research in Manitoba indicates that ducks intending to nest in an area scatter almost immediately upon arrival, whereas those only stopping briefly remain in close-knit flocks.

Another factor, this one behavioural, strongly influences duck distribution during the summer. When the drakes begin their moult in late June, many of them leave the ponds to gather on lakes. Here they skulk among dense bulrushes and cattails or crowd into large offshore flocks, especially the diving ducks, during the two to four weeks that the complete loss of their flight feathers renders them vulnerable. However, hens with young remain on the ponds until their families are fledged, since adult females do not moult into flightlessness until August.

We walk along a track leading to a low-lying meadow at the north end of a sizeable pond. It is empty of bison; apparently empty of all animal life except a queen bumblebee which drones across our feet as we leave the trees. It is early afternoon and the full sun floods the meadow. On the far side the heated air ripples up in transparent waves that shimmer against the tree trunks. A light breeze puffs warm air against our skin. Underfoot the old grass is dry and brittle, the new softly yielding, so that our boots often crush the young tissue, releasing its fresh spinach smell.

I leave the track to follow along the west side of the pond. A few strides, and I look down to step over a log. Just beyond, head resting on its folded legs, lies a bison calf. It is tiny, not more than a day or two old; its thick, shaggy fur is a bright yellow-tan. Then its large, limpid eyes open, and it raises its head slightly looking at me vaguely, uncomprehendingly. Fortunately it is too young to react with fear. We back slowly, turn, and walk steadily the way we came. The nearest big trees are a hundred yards away, but we know the calf's mother is not in that direction, or at least was not a few minutes ago. Nevertheless, all the way my heart thumps rapidly, heavily, so that I feel its thrust all through my body. I am suddenly aware of the map of my arteries; my skin sweats damply under my clothing. We will have little hope of bluffing or diverting the cow should she charge red-eyed with rage out of fear for her newborn's safety. Then the trees are closer than the calf, and we have seen nothing of the cow. Soon we lean weakly against their smooth, hard, reassuring safety, and scan the meadow behind carefully. Still no sign of the cow. But an abandoned calf would hardly have been calmly resting, and we are certain its mother is somewhere nearby.

In a few minutes we make our way noisily through dense forest along the pond's east shore and spend nearly three hours squinting into the sun, identifying and counting ducks, grebes, Franklin's gulls, and the first greater yellowlegs. Several times the pair ring the pond in flight uttering loud, belling cries.

During the afternoon a little flock of warblers, the first, appears in the poplars and willows near the shore, feeding as they trend northward. Except for two orange-crowned warblers—dull, olive-yellow birds whose orange feathers show only when raised in excitement—they are all myrtles. As they flutter to catch insects their lemon-yellow rump, crown, and side patches flash in a setting of grey and black, for they are males, the vanguard of the migrants. The movement and colour initially drew my attention to the birds, since they converse only in low, indistinct contact notes.

It is well over a week before I return to the vicinity of the meadow and pond.

The flush of heat for the past three days has been almost oppressive. The air and sun are suddenly mid-summer while all things earthbound pause on the threshold of spring. Overnight the willows bloom in a golden haze of ripe catkins that hum with flies and bumblebees. Coltsfoot stems, now several inches high, hold up bouquets of papery white flowers. Leaves have appeared with the marsh marigolds. Many aspen, saskatoon, gooseberry, currant, and rose buds are green-tipped. Buffaloberry flowers cast delicate sprays of yellow against the bare, grey-brown twigs. Thick red balsam poplar catkins emerge like paste extruded from a tube. But I have to look closely at the hazel to find their pollen-dusted catkins and red-tentacled female flowers. On the ground the matted leaves are speared in dozens of places with the first growth of perennials—vivid, petit-point green on a multi-brown background. And then the scarlet funnels of a cup fungus, only an inch and a half high. For a while now plants will seem close kin to animals, not the quicksilver of insect, bird, or mammal, but the languorous stretching and gliding of coral polyp, starfish, or clam.

By early afternoon the heat is too much for me and I seek the only shade, under spruce. The interlocked foliage reflects heat as well as light, and the interior of the stand is pleasantly cool. I drowse for a little and cleanse my lungs in the resinous air. Released by the kindling sun outside, the fragrance permeates the whole stand, yet I will be able to detect it only a short distance into the poplar forest. Dense spruce creates its own special environment. It is to the surrounding forest as an island is to the lapping ocean. Many of the plants and animals associated with these conifers shun the poplar, and vice versa. I am aware of birds passing and calling beyond the spruce island, but they either go around or above it. Within are only a pair of husky-voiced boreal chickadees searching the branches for food; the thin, nasal piping of a red-breasted nuthatch out of sight in the dimness above; and the methodical tapping of a downy woodpecker.

I rouse to a red squirrel descending a trunk head first in scratchy jerks. On the ground it merges with the rust-brown, deep-layered needles. The squirrel moves slowly, nosing, and sometimes digging a little with its front paws. From time to time it sits erect and looks about, or chews on some edible it has picked up. Then it disappears into a tunnel angling down between the raised roots of an old spruce. A minute or so later it emerges at another hole, one of many from the underground labyrinth. A ripple of bounds takes the squirrel almost to the base of a tall tree. In one leap it is two feet up the trunk, from where it hitches up until it reaches the level of the stout branches. It bounds easily along one of these; pauses; makes a long, flying jump and catches the end of a branch on the next tree. Squirrel and branch sway wildly for a few moments, but then the squirrel is on its way again, at ease along an obviously familiar route. When I leave the monastery of the spruce it is still hot, and the sun glares down from a pale sky scattered with little, bleached clouds.

Sunset brings relief, although the air stays warm for a long time after. Bird song and frog chorus replace the brilliance of the light. Then, out of dusk dimming to darkness comes the woodwind trill of a toad. The song is a monotone of constant volume that continues four or five seconds, then abruptly stops. It is repeated at frequent, but irregular intervals, and it is while I am waiting for the next trill that I drift asleep never hearing it.

From the first moment of its appearance the sun burns hotly in a jaded sky strung with wisps of cirrus. During the past days the air has sponged all surface moisture from the ground. Dead leaves are dry-crisp again. Without rain plant growth has slowed.

The breeze which was south at sunrise, now near noon, strengthens and gusts in from the southwest. Short, steep waves are building on the lake and beginning to break in white. I find the wind irritating. It carries no coolness with its rushing onwards, nor does it hold to a constant direction. Everything is in motion. The pole-thin poplars oscillate like pendulums, but each with its own amplitude and timing. For once the forest seems to be no more than a collection of trees, a crowd of strangers. The interlacing crown branches clack and click and slap against each other. Two trees in contact screak as the living sways and the dead, reclining against it, shifts a little. Diminished, but still restless and probing, the wind reaches down to the ground. Shrubs flex under its pressure; dead leaves stir; the old, shrivelled grasses, asters, and peavine whisper their resignation.

Nor have the ponds been left in peace. They are dark-ruffled with wavelets fleeing the wind. The great beds of tall cattails, strung taut, yield a concoction of rustles, snaps and sighs which wax and wane with the surging of each gust. A pair of red-necked grebes keep close to their nest anchored to a sedge clump. Near the centre a black and white bufflehead bobs on the water, facing the wind. His chubby proportions and seemingly contrived pattern give him the manufactured look of a bathtub toy. But when he dives, coming

to life, the illusion vanishes. A pair of mallards, three drake widgeons, and a pair of blue-winged teal are spaced along the far shore feeding. All head into the wind. The red-wings are deep down in shelter, all but one who skids briefly across the tops of the cattails before dropping out of sight.

The wind blows harder. The lake is now a seethe of white and sullen blue, but in the telescoped perspective of my viewpoint a few feet above the water the distant waves are deceptively small. Several ring-billed gulls are in the air playing deftly in the tumbling currents with the grace and skill unique to gulls, although most other waterbirds have sought lee shelters.

The waves roll into the shallows to topple in a crush and sparkle of foam. The working of the water, back and forth, up and down, swish and swirl has stirred up the bottom and muddied the lake for some distance out. Bits and pieces float on the surface, drifting slowly to a new landfall: sections of root and branch which must have been going the rounds for many years—barkless, sanded smooth, weathered grey; carunculated lengths of cattail tuber; a discarded plastic bottle, turquoise and alien; leaves; strands of recently sprouted pondweed; a half-plucked cattail seed head.

The too-bright sun dims slightly and I look up at the sky that has been an inconsequential backdrop for the past several hours. It leaps forward, bigger and now more important than anything else. Down along the horizon it is dirty, brown-smudged with topsoil that had dried to dust in ploughed fields. Over the rest the cirrus streamers, interlayered with opaque sheets of high stratus, curl out across the bubble of blue. In half an hour the stratus thickens, masking the sun. To the west, low down still, thunderous clouds are massing. They lumber forwards spreading their gloom. Where it is not breaking the lake is darker than the sky. It appears mineralized, like a huge outcrop of gabbro laced with milky feldspar.

The storm front is a sheer, purple-tinted wall behind the forest. A distant wire of lightning snickers down to earth. Fifteen seconds later thunder rumbles above the sound of the wind. Now that the storm is close it seems to rush at me. The dark cloud reaches for the far horizon. The playing gulls have gone. I make for the house. The wind crashes through the trees and rolls over the lake; leaves are sucked up into eddies; poplar catkins are torn loose and hurled away. The lightning increases—from cloud to cloud, cloud to ground in searing, white-hot strikes, and fluorescent, reflective sheets. The atmosphere has darkened to dusk. A blast of cold air slams down, momentarily flattening the waves. Then it flows outward and joins with the wind.

The writhing clouds are overhead, annihilated by the plunging rain. The far shore, then the islands dissolve in grey mist. Lightning flares almost continuously; the echoing thunder cracks, roars, grumbles in ceaseless dissonance. I can see the wind, but not hear it.

In slightly less than an hour the light brightens a little, but heavy, layered clouds ceiling the entire sky. The rain ebbs somewhat. The wind drops to a strong breeze which no longer strains the trees, and raises whitecaps only in the middle of the lake. It has veered farther, and now comes out of the northwest.

The storm has upset my sense of time, and I do not realize the day is nearly over until real dusk draws up the cloak of night.

The morning struggles awake, slowly, coldly, wetly. There is little birdsong; the grebes are subdued; frogs and toads silent. Tree swallows flutter around the spruce where flying insects have taken refuge in the dry, dead-air spaces between the shingled limbs.

It is a day to wander through the dripping forest doing nothing in particular, looking for nothing, touching nothing, absorbing all. The melancholy of the hours is soothing, a transient episode that makes no demands on the senses, muscles, or mind.

The rain stops by mid-afternoon. Later, rents in the overcast open on pale, transparent blue. Just before sunset the cloud lifts off the horizon. A flush of golden light glazes everything it washes over, but it is too late for real warmth, only its imitation in colour.

The Arctic-cooled air is crisp to the point of freezing. Eastward the remnant clouds are ablaze with sun-fire. The colour reels from dusty rose through glowing magenta, vermilion, burning gold, to the whitened glare of iron in a forge. The sun shimmers behind the distant trees, sliding smoothly into view. For a few seconds I can feel the titanic roll of the earth toward the sun. Yet once the orb is free in the sky the sensation passes, not to be recaptured. Soon sun-warmth touches my hands and face although the surrounding air stays cold.

Bird song quickens. The day surges alive. At the beaver lodge near the end of the point a magpie is bathing. Balancing carefully it sidles down an outleaning branch into the water. With its belly just touching the surface it dips its beak and head showering water over its body with a quick, backward motion. This is repeated several times; then the bird cups its wings and splashes more water over its rump and long, trailing tail. Bedraggled, the magpie climbs out of the water. Like an umbrella opening, all the body feathers are fluffed out, the wings droop, the tail sags. The magpie shakes vigorously, then preens using its heavy, coarse bill with delicacy and precision. From time to time it reaches back to the base of the tail to smear oil from a gland, first to its beak, then to its feathers. After a last dowdy shake it sleeks its plumage and restores black and white formality. Just before the magpie flies a pair of mallards swim around the point. The drake's sunlit head gleams with a metallic green lustre. He turns to his leaf-brown mate, changing the angle, and is crowned in royal purple.

At this moment a merlin, blue-grey above and dark-streaked white below, slides over the point on set wings. Much smaller than the mallards, he brings no fear to them. Then his long, falcon wings flicker in shallow, rapid beats as he grows small across the bay, to be lost instantly against the grey and white trees.

On the way home for breakfast I see a white-crowned sparrow hopping about on the wet leaves. So contrasting is the black and white striped head that it seems to be an independent entity, an oversized beetle, or perhaps a moth.

The morning warms to sun-basking day, but not every place equally. I breathe the cool-scented exhalation from spruce as I pass a clustering of young trees. Shadow-barred poplar forest is lukewarm except in hollows where the chill has ponded like water. Damp with humidity, the heated air over meadows and shrub slopes diffuses upward redolent with the mingled perfume of earth and living plants. Willow-studded fens are catchments of stilled heat, cold and wet underfoot.

More birds have arrived. I hear several white-throated sparrows along a mile of forest trail, but glimpse only one as it darts off through the undergrowth. Their song is an ethereal fluting which, after one or two long notes, usually ascends in four simple, measured steps, the last two quavering slightly.

Where low ground at the head of a bay is touseled with the sun-bleached hair of sedges and grasses, a hidden LeConte's sparrow delivers his brief insect trill. I wait hoping he will move up to some visible perch, but he does not, although he continues to sing repeatedly.

Six cowbirds fly into an aspen at the edge of an upland meadow. Five are glossy black males with brown heads, one a dull, mouse-grey female. They all perch close together and the males begin to display. Their movements are highly ritualized: the beak is pointed skyward; then all the body feathers are ruffled, and I see the throat ripple as the song is squeezed out, a soft gurgling that ends in a high, nasal *sneeeeee*. The final phase is one smooth movement that is almost too quick to follow. It begins with the neck being arched, the tail fanned, and blends into a deep forward bow. Simultaneously the wings are spread out fully and the tail tilted up. At the end of the bow the wings snap shut and the feathers slowly subside. To finish, the beak is wiped against the branch. The entire display takes only a few seconds, if that. The males court the female in this way for some time with no apparent response from her. Suddenly she leaves in swift, direct flight along the edge of the meadow. The males hurry in her wake.

I hear the sewing-machine trill of a chipping sparrow coming from high up in a poplar. I eventually locate the trim little bird with its distinctive brick-red crown. Each time he sings a tremor vibrates his whole body; sound becomes visible.

Were birds less vocal I am sure many would be overlooked, and bird-watching would not be the popular hobby it is. Winters here are profoundly silent primarily because there are so few birds. Again I pick out another new arrival first by voice. From the shrubs and saplings at the forest edge comes the ascending three-part buzz of a clay-colored sparrow, but the cryptic, dun and off-white bird is even more difficult to find than was the chipping sparrow. I hear and see several more during the day, all in similar shrubby habitat.

In the afternoon I approach a large beaver pond along an old forest track. Disturbed flies jet from shrubs and the ground. Butterflies have responded to the sun: tortoiseshells, mourning cloaks, white admirals, angle-wings and, newly emerged from chrysalides, silvery blues. The last are dainty butterflies with a wingspan of barely an inch.

Ahead a mourning cloak rests on the leafy ground in the sun, its wings spread wide. A second arrives, prompting the first to fly up. Both ascend erratically, the first just behind the second, until they top the tree crowns and disappear. About a minute later one returns and lands at the same place. Soon another mourning cloak flutters down and the pursuit flight is repeated. Again a single butterfly returns to the original site, but on this occasion no challenger attempts to usurp it. I cannot tell which individual was successful in this interchange, but perhaps the resident, which may be a territorial male.

A patch of buffaloberry in flower draws a variety of insects. Several black-spotted ladybird beetles clamber slowly over the blossoms. Wasp-striped hover flies, after a few moments' suspension on speed-dissolved wings, land to sample nectar and pollen. Often one remains long enough to test every flower in a cluster. A bee fly arrives and hovers, probing for nectar with its straight, wire-thin proboscis. When it leaves it skims along close to the ground. It is a large fly with a swollen abdomen and its whole body is thickly furred with long, brown hairs. To me it looks only vaguely bee-like. It is in its juvenile stages that it is intimately associated with real bees. The eggs are laid at the entrance to a solitary bee's nest. On hatching, the larvae quickly wriggle down the tunnel and burrow into a cell. For a while they may eat the stored honey and pollen, but eventually transform to become an internal parasite of the developing bee.

Threads of light reveal the newly woven web of an orb spider. I find her at the centre, head down, legs drawn in, her bulbous sepia and orange marked abdomen resembling an exotic seed pod.

Once I almost step on a queen bumblebee crawling on the ground in search of a nest site.

Many herbaceous leaves have grown enough to name: violet, bluebells, snakeroot, peavine, sarsaparilla, fireweed, baneberry, fairy bells, aster, and mayflower. They possess a pristine perfection that will be lost in a week or two as the tissues mature, the green deepens, and nicks and holes are cut out by feeding insects. Mayflower leaves emerge from the ground tightly furled and do not unfold until the whole blade is clear; the tightly compressed, bronzed sarsaparilla leaflets top slender naked stems like tiny cupolas. Most of the others are more ordinary in appearance, but all contrast vividly with the blemished, year-old green of bishop's cap, wintergreen, and twinflower.

While most plants are only now feeling their way into a new season, flowering is over for aspen, hazel, and alder. Many of the male aspen catkins have fallen and lie in drifts on the ground or draped limply over shrubs. The female catkins have lost their silvery hairs and the enlarging seed capsules are green. On the other hand, balsam poplar is just nearing its peak. Every sizeable tree is decorated with clusters of red catkins that dangle like Christmas tree ornaments. In one a red squirrel is feasting, turning and nib-

bling catkins held in both front paws as we would eat corn on the cob.

I come out of the trees at the top of the hill slowly, half-crouched, the telescope already set in focus across the pond where, in a stand of drowned poplar, the eleven nests of the great blue heron colony are clustered. Nine nests are in good repair; herons are at seven of them. Two birds are sitting, long slender necks and small heads disproportionate without the counterbalance of the legs. Presumably they are laying, or perhaps already incubating. Other herons, ten in all, stand on nests or on nearby branches. I watch for several minutes, but nothing much happens. The birds look about, preen and stretch. Just as I am diverted by a yodeling lesser yellowlegs, a heron flies in from the west. Those at the colony watch its approach and landing without comment. As the bird descends, its long legs reach out and the wings back stroke. It seems to be all appendages with the body little more than a point of attachment. Then it is at rest, neck curved gracefully, broad wings folded closely to its sides.

Around a corner of the pond, cut off by the swell of the hill and the trees on its crest, a mob of Franklin's gulls cries in near hysteria. More are flying in. I walk along the edge of the forest until I can see into the bay. Over a hundred gulls, including a few Bonaparte's, are on the water twirling, darting and stabbing their beaks at the surface. Hordes of adult midges are emerging from their aquatic pupae and the gulls are gorging themselves. Yet many midges escape, and with binoculars I can see them tower into the air. There gulls in flight swoop and twist, snapping up hundreds more, one at a time. Now in complete and fresh breeding plumage the Franklin's gulls show a faint rosy blush underneath. The colour seems to come from deep within the feathers, rather than being a fixed, surface pigment, and it deepens in shadow. Direct sunlight drives it out of sight, down to the birds' skin. The gulls continue to stream in until more than three hundred excited, seething birds have gathered, but their frenetic emotion is not contagious beyond the limits of the flock. Ducks feeding around the pond's edge, a pair of displaying red-necked grebes near the beaver lodge, and birds in the surrounding forest remain unaffected.

After a night of spattering showers heard dimly in dreaming sleep, I step out into a morning bright with golden sunshine and jewelled rain drops. Frayed clouds drift eastward leaving the sky clean-swept behind. The night's chill is quickly dispelled. Bird voices overlay one another, weaving a richly textured macramé of sound, yet one in which most of the individual yarns are still discernable. The frog chorus is an unceasing accompaniment here by the lake, but each day progressively fewer wood frogs participate. Their mating and egg laying mission to water is short, and once accomplished, most adults make their way back to damp forest.

To take for granted that leaves appear, expand, and finally give green solidity to the summer forest is to bypass a brief but exquisite season. Already I am almost too late for the first leaves of currant and gooseberry now spread flat as bright emerald salvers. But eventually I find a few still clasped at the base by the bud scales, the pleated blade forming a funnel scarcely half an inch long. The leaf is red-margined, in common with most very young leaves whose chlorophyll is not fully developed. Cherry and saskatoon leaves emerge folded flat along the mid-vein, and flank the longer, tightly budded flower spikes. The former are smooth, shining and bronze; the latter, silky-hairy and greyed-green. Red-osier dogwood stems are each tipped with pairs of amber flame points curved around a sphere of beaded flower buds. What I had taken for bud scales on the buffaloberry now reveal themselves to be diminutive leaves opening from their winter-long clasp like pages of a book. The deep jade-green of the upper surface is surprising. Today other shrubs have advanced little beyond tentative green points, but in a day or two if it continues warm all leaves will be out of their brown wombs.

As I follow a forest trail a mourning cloak alights almost in front of me. I stop. It spreads its wings open and is still. Their edges are tattered, the marginal bands bleached almost white. Its life seems nearly done, but if a female, it will have by now deposited eggs on willow and poplar twigs. There is no biological need for it to go on living; the next generation will continue the species. I step forward, and the butterfly flutters away without difficulty, its worn appearance belying its condition.

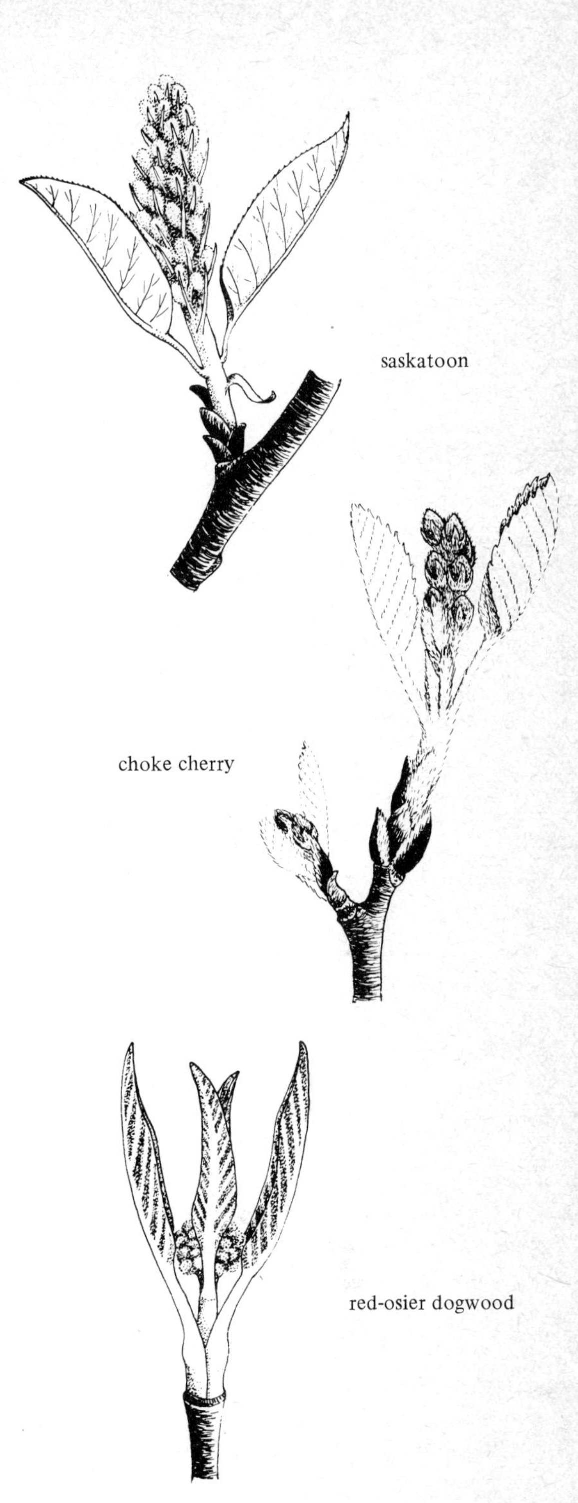
saskatoon

choke cherry

red-osier dogwood

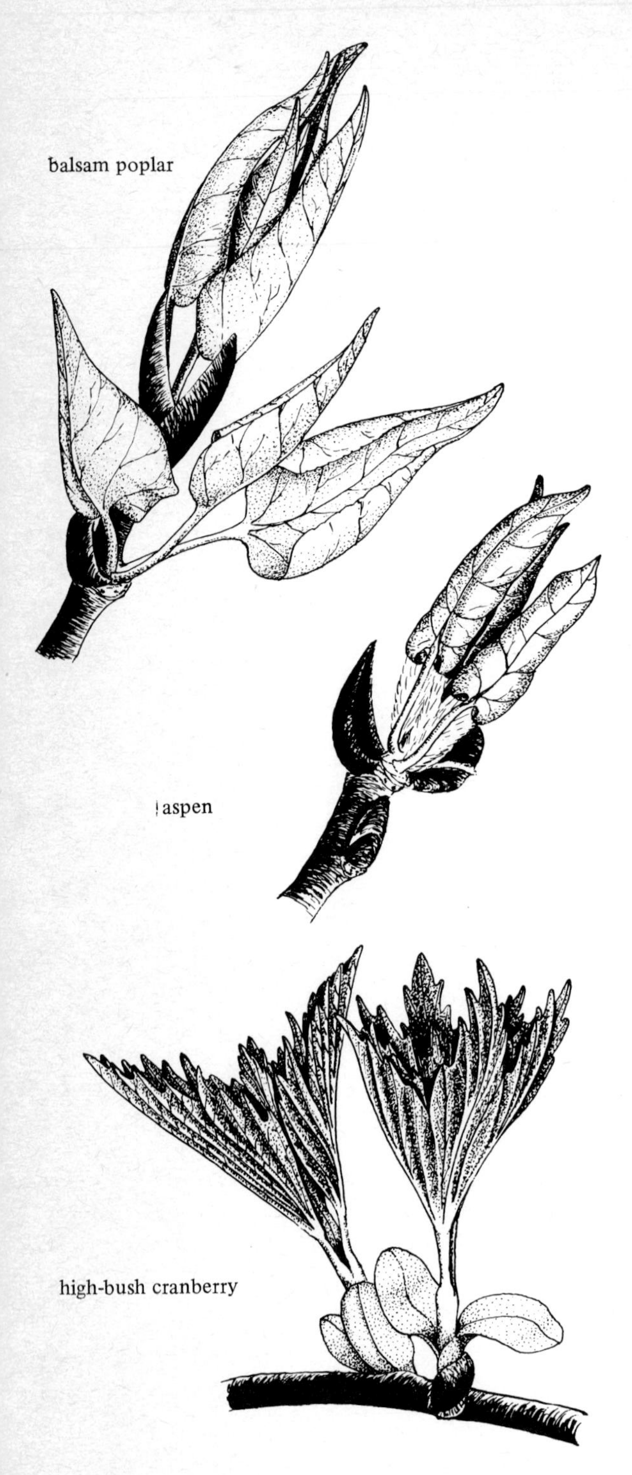

On a slope an angle-wing flits close to the ground. It seems to be searching for something not to be found, as it often alights, yet momentarily; then is off again on its meandering quest.

From somewhere, it is impossible to tell exactly how far or more than the general direction, a mourning dove's low cooing flows around me as if I were a rock in a stream. The sound is not so much sad as infinitely soothing, a balm to all anxiety.

Two days later flecks of green now shine everywhere, although at any distance are lost in the pervading browns and greys. It is a strange sensation, and I imagine myself in the interior of a vast algal colony, the near leaves it constituent cells.

The birth of foliage follows distinct patterns. Poplar leaves emerge in clusters, but on aspen each blade is tightly curled in on itself, while on balsam each is wrapped successively around another. Birch leaves, all bright, yellow-green, are funneled and pleated along the parallel lateral veins. High-bush cranberries push out erect fountains of green that only gradually cascade outwards. In contrast, low-bush cranberries almost immediately turn their purple, three-lobed palms horizontally. Sharply toothed hazel leaves, glistening in their newness, are crinkled and pebbly in texture. I move on seeking others. Rose leaves are furled like umbrellas, only here and there a paper-thin leaflet bending free. At first the variety seems infinite, but soon I find themes repeated, although each species produces a colour, surface texture, and shape that is never exactly duplicated.

Bud scales drop off unnoticed except for the sticky ones from balsam poplar which coat my boot soles by the end of the day. Do small mammals and ground birds also find them a nuisance?

With my attention focused on plants I am oblivious to all else until an oriole proclaims his arrival with a commanding three-note whistle. Despite the power of the song its tone is pure. I look up and find the bird moving restlessly from branch to branch, the sun firing his jig-saw black and deep orange, coal and flame. Now I hear other birds again, and a chipmunk somewhere nearby repeating its *chup* of warning with mechanical regularity.

Late in the afternoon I sit on a log near the end of the point. The beavers are out and swimming at the surface. One adult rounds the point and disappears. Two others, yearlings to judge from their size, keep close to the lodge diving often in play. From time to time I hear their groaning murmurs. Close to shore a pair of mallards and a drake blue-winged teal feed silently, skimming the surface with their beaks.

With three hurried *weep*'s a dark shorebird veers, then checks with wings raised high as it drops lightly at the water's edge. For a moment it is frozen, then its tail bobs up and down rhythmically as it walks about dipping its beak into the water. From the side the black-marked white underparts seem unrelated to the mud-brown head, back and wings; from above, the bird would be one with the shore. The spotted sandpiper feeds for perhaps two or three minutes. Suddenly it leaves. One instant the bird is motionless, the next receding low over the water, wings alternately vibrating and rigidly set, its high call notes coming back to me from afar.

As the sun nears the horizon ring-billed and Franklin's gulls fly in from the north and west where they have probably spent the day foraging on farmland. They are quiet and leisurely, in small groups, as they lower over the lake. However, as the night-roosting flock grows, conversation increases. Behind, in forest, the attenuated songs of the first Swainson's thrushes ascend. Each begins with a long cue note, then breaks into rolling phrases that thin away to nothing. For a little while I continue to sit shrouded in sound. From an orchestration of pervasive frog chorus and grebe braying other vocalists surface intermittently as divertimenti—toads trilling, a snipe winnowing, an invisible spotted sandpiper calling plaintively, a few robins singing snatches of song between expulsions of call notes, red-wings whistling and scolding, gulls keening, ducks muttering and quacking. Near dusk a single, poignant white-throated sparrow begins fluting; and once, on the far side of the lake, coyotes yip briefly. Ultimately, darkness brings near repose.

For the past several days I have woken at first light and dressed immediately, caught up in the urgency of the too short summer when every sunlit hour overflows with life. Already it is past the middle of May.

A quick snack and I step out the door as the just-risen sun paints the landscape in surrealist hues. Wherever I look birds are moving —feeding, courting, collecting nesting material, chasing intruders. Their songs and calls now so interweave that often it is difficult to separate individuals. But more migrants are to come, and I am alert to new songs. The familiar vigor of a singing yellow warbler draws me to a dense, lakeside willow thicket. He is not alone, although the first bird I see, well above the ground and moving from twig to twig, flashing brightly. Then I become aware of other warblers flitting more discreetly through the tangled stems: a couple of magnolias, black and white above, all yellow below; several myrtles, similar, but with mere flecks of yellow; a palm, rich earth-brown above with streaking of the same on his pale yellow underparts; a Wilson's, his jet black crown sitting on plain yellow like a separate cap. Yet another bird moves, and I recognize an orange-crowned warbler. A Tennessee, trim but subdued in grey, white and olive, mounts to the upper twigs and delivers a long and emphatic series of musical *chip*'s on three different levels, accelerating toward the end. But aside from the yellow warbler he is the only one to sing, although a constant patter of call notes comes from the others. The warblers slowly move through the willows away from me. Then a redstart flies, flaring orange on his wings and tail.

I turn back into the greening forest. Red currants are hung with sprays of dusty-rose blossoms, gooseberries splashed with green and ticked with the white of scattered buds. The stems of most herbs are now several inches high and flower buds are forming on fairy bells, mayflowers, baneberries, bishop's caps, strawberries, and white violets. The old bunchberry leaves have completely collapsed, but the leaf buds are opening. I see a sarsaparilla which has pierced a dead hazel leaf, raising it aloft and at the same time imprisoning its own leaves. Their points are held together in the hole, but underneath swell in a grotesque bulge. Doubtless in a few days it would free itself, but I hesitate only momentarily before lifting off the hazel leaf. Almost at once the released leaves begin spreading open.

As I bend to contemplate the elegant curves of an emerging mayflower leaf, a sepia and ochre morel detaches from its background of sepia and ochre leaves.

I walk on. The trees are full of least flycatchers, drab brown birds whose hiccuping *che-bec* will be an integral forest element until mid-July. A white-throated sparrow starts up from the ground where it has been feeding, pauses briefly in a hazel, then flies a long way. Two orioles sing overhead; one chases the other, and the birds arc away in firey orange. The trees here are mature and widely spaced; many have died. Some of the dead have blown down, others are headless, mouldering stubs. A newly arrived house wren bubbles his enthusiasm, for somewhere will be a cavity to fill with a nest. He sings now, for a mate and to establish title to a territory.

The new song is almost a robin's, but when I find the bird on a high, leafy branch he presents a white front cravated with geranium. The rose-breasted grosbeak sings many times without changing position. Then he flies to another tree, revealing only then the pink flush of his underwing linings.

I pause at a flooded willow and alder fen near the lake. Most of the male willow catkins have fallen, and the swelling seed catkins are half hidden by young leaves. Below, the coltsfoot flowers are over and the lance-headed leaves clear of the water. Marsh marigolds are handfuls of deep yellow saucers, as glossy as if varnished. Just before going on I look out across the wetland again. Something moves and binoculars put a snipe in my hand. It is preoccupied with feeding and has not noticed me. It plunges its beak into the soft peat, pauses to swallow, plunges again. Its activity is so concentrated that, at times, it seems only the pumping head and neck are alive. At long intervals the bird does step ahead, but not until it has probed every point of the semicircle within reach. Gradually I retreat without disturbing the snipe, and only then become conscious of the mosquitoes whining around me.

After lunch I drive to Tawayik Lake, walking the last half mile. The poplars here, although well spaced, are slender and somewhat stunted, and shrubby undergrowth is much reduced. This is one of the few places where conditions are satisfactory for ovenbirds, large ground-dwelling warblers, and a loud, declarative, *teacha-teacha-teacha-teach* signals the return of at least one male. The ovenbird is persistent and reiterates his song many times before I leave the trees.

Overhead a solitary vireo moves slowly through the crown branches. Like his movements, his short slurred song is lethargic, with long pauses between the grouped phrases.

Light shines beyond the trees as I near the lake, and I walk more quickly. Suddenly a small bird flutters almost from my feet, alights near the base of a tree, and begins to creep up the rough bark. It is a black-and-white warbler, another species which has deviated from the typical perching and foliage-feeding behavior of most wood warblers. This one mimics the real tree creeper in feeding, taking insects from bark crevices, and to some extent in plumage, being similarly streaked, although boldly. Nevertheless, in general appearance it is still a typical warbler.

Sun pours into the forest. In its warmth a few worn overwintered butterflies wander through the undergrowth. Most will soon die, and it will be July before any numbers of fresh adults are flying again.

I stop behind a screen of saplings, cherries, and saskatoons that fringe the forest. Ahead, the grassy meadow reaches to the lake. Wind I have barely noticed until now combs the long grasses, old and new, this way and that. A staggered line of aspen behind the shoreline levee and scattered groves in the meadow stand fast as they have done already for thirty or forty years. A small herd of bison, including several cows each with a young calf, grazes far enough away to be of no immediate concern. I push into the open. Sun and wind engulf me.

I walk slowly to the lake. A male bluebird is perched high up on a dead branch. Every few seconds he darts out a few feet, snaps up some flying insect, returns to the same perch. I hear the beginning to his gentle, warbled song once, then a gust overwhelms it. I disturb a second bird from a like perch in another tree, and assume it to be the bluebird's mate. Instead, it is a Say's phoebe. For a northern flycatcher it is showy, dark grey-brown upper parts heightening the contrast of the rusty-orange belly and black tail. But like the familiar eastern phoebe it pumps its tail up and down while perched, and like it also adopts buildings for nesting. Will this one stay? There are no definite nesting records for the park, and the eastern phoebes who arrive much earlier seem able to hold their own, if indeed they are ever challenged by the occasional Say's phoebe. I observe the phoebe for several minutes while it, like the bluebird, makes graceful sallies after insects. It has the additional appeal of the infrequent visitor. The phoebe eventually flies to another tree and I go on.

From the grass ahead a savannah sparrow starts up, then drops out of sight in a patch of buckbrush and rose. The simple, desiccated songs of clay-colored and chipping sparrows whirl on the wind.

As I near the lake its flat, reflecting face widens from a slit to be finally all the space between me and the horizon where a narrow band of forest distinguishes it from the sky. Ducks and grebes litter the water, close in and far out, in apparent flocks and scattered pairs, in total probably five or six hundred individuals. The light and waves make distant identification difficult, and I concentrate on birds east and west of where I stand.

Two pairs of red-necked grebes resolve a dispute. Two of the birds who had been circling each other swim apart, stiff-necked, beaks tilted up, tails cocked, and rejoin their mates. Then the pairs glide farther apart, the grebes of each side by side, heads high and braying in unison. On this occasion their differences have been settled by the parading of threats.

In deeper water several eared grebes dive repeatedly, food their interest of the moment.

A ruddy duck displays to his mate. Throughout, his fanned black tail is held stiffly erect. Raising his black and white head high over his puffed out chest he begins a rapid, vertical jackhammering which lasts for a few seconds, and ends with his head sunk down on his breast, bright blue beak resting on a pillow of rufous feathers. But I am too far away to hear the accompanying grunting clucks.

Shovelers swim in the shallows mouthing the water with their broad spatulate beaks and sifting out small edibles from debris. The green-headed drakes look, at first glance like misaligned mallards with their white chests and russet sides, instead of the reverse.

A group of dumpy shorebirds feeds on a scallop of mud towards the sun, jabbing their beaks again and again into the ooze. They are dowitchers and resemble heavy-bodied snipe, but are darker and reddish underneath.

A little beyond them, on open shore trampled by bison, a killdeer picks it way over the uneven surface. Its precise patterning gives it a tailored appearance in contrast to the mottled and striped plumage of the dowitchers. Yet it is not conspicuous, and only movement makes it readily visible.

All this time, over the water and the sedge beds in the northwest corner of the lake I have been aware of black terns hovering, dipping, rising, side-slipping, feeding and courting, crying thin, reedy notes. In this intense light they are decisively two toned: grey upper parts faded, black underparts deep in shadow. Tree swallows weave swiftly among the slower terns, by comparison making them seem almost clumsy; although against gulls terns are like swallows.

I step back from the water and walk west along an entrenched bison trail. The herd which was here when I arrived has moved farther away. I study them from a hundred yards or so, and occasionally glance around in case another band should move in. The

lush summer meadows around the Tawayik Lakes draw and concentrate bison. Wallows used and abandoned pock the grass; bark on many trees has been rubbed smooth; tufts of underfur drape the shrubs; and cakes of dung, dry, age, and disintegrate in the sun, wind and rain. All the cows with new calves graze without pause, although several yearlings and two-year-olds are lying down chewing cud. A magpie planes down, lands on the back of a cow, and stays for perhaps a minute, its movements suggestive of snapping up flies. She doubtless feels the bird's weight, but ignores what may be a common occurrence. Possibly the attentions bring relief, for a bison's short tail is an ineffective fly switch. Small birds are about the bisons' feet, flying up and dropping down into the grass again. They must be cowbirds, restored now to their aboriginal role.

I walk slowly, but nevertheless put up the dowitchers, the killdeer, and a spotted sandpiper I had not seen. All protest the disturbance.

At the edge of an old wallow, not yet grassed over, clumps of golden corydalis are in full bloom. The thin, divided leaves and finely structured, tubular flowers give the whole plant an aura of delicacy, yet it thrives on exposed, bare ground where other plants do not compete with, or shade it. Inevitably dandelions grow here too, crude and garish beside the corydalis. In the grass masses of blue violets are almost overcome, but the white-starred strawberry flowers sparkle like the highlights on waves.

I return inland, close to the margin of the encroaching forest. Brushing through some shrubs I raise a cloud of green lacewings. They are beautiful but insubstantial, translucent beings who seem barely able to sustain flight. It is difficult to associate them with their grub-like larvae who have an insatiable appetite for aphids. The adults also eat aphids, but less voraciously. Already their food supplies are multiplying, for today I have found a number of small aphid colonies on the young leaves and shoots of poplars, dogwoods, and willows. Some of the lacewings land again, their golden pinhead eyes gleaming in the sun.

Out in the open cabbage white, saepiolus blue, and silvery blue butterflies float and flutter like detached strawberry and violet flowers. A few dragonflies dart among them hunting midges and mosquitoes. They are skimmers, with clear wings and a deep claret body. Suddenly a tiger swallowtail comes out of the trees, a flamboyant butterfly that might have been transposed from the tropics. With its four to five inch wingspread the swallowtail is our largest butterfly by far. Its size and flashy black and yellow striping accurately hint of its tropical connections, for most of the family's species range in those latitudes. However, one, the smallest, is restricted to the north, its farthest limit almost at the Arctic circle in Yukon and eastern Alaska. The swallowtail dips down over some shrubs, then across the meadow, and finally out over the lake. It angles into the wind flying strongly and with a determination most butterflies lack. Before I reach my starting point I see three more, all recently emerged from overwintering chrysalides.

On the way home I stop and walk through a tract of mixedwood forest on hilly ground south of Astotin Lake. White birch catkins ripen as the leaves unfold, and when I touch some low-hanging ones my finger comes away with a patina of yellow pollen. Where a fallen tree has opened the forest to the sky a patch of fairy bells is in bloom, and a little farther on, near the bottom of a slope under spruce, a few kidney-leaved violets and skunk currants are beginning to flower. All three are white.

Marsh marigolds glow beneath willows. I move closer—a rush of splashing and snapping twigs as a cow moose lunges away with a hollow grunt, followed closely by her spindly, reddish calf. Had she stayed still I would never have seen her; had she charged I could not have eluded her. Our relief is immeasurable when large animals who could maul us flee instead, yet we are often bewildered by the secrecy and timidity of the many smaller ones whose confidence we would welcome, but who in turn find our gigantic size threatening.

Back through the spruce where they begin to mingle with birch and aspen, a muted brown bird flies to a low branch with a soft *tuck* of warning. The grey-cheeked thrush stays for several moments, glancing about for the cause of its alarm, but it will not stop to sing, mate, and nest until it is deep into the northern boreal forest. It flies off through the trees not far above the ground, a dead leaf spun by the wind.

By late afternoon I am at Astotin Lake. A kingfisher rattles and flies from a dead branch over the water to another dead branch farther along the shore. There it peers down, but apparently no fish rise close to the surface. It soon flies again, calling as it goes.

On a half-submerged poplar trunk angling out from shore a solitary sandpiper steps carefully, dipping its beak often into the water. It is an attractive bird, white underparts softly marked with grey-brown streaks along the sides, dark upper parts flecked with white, and narrow white spectacles around the eyes. Legs and beak are blackish-green. From time to time it bobs gracefully like the spotted sandpiper, but without its exaggeration and frequency.

Over by Long Island several white birds weave back and forth in shallow parabolas, now and then touching the water. I hardly need binoculars to verify that common terns are back. Between the island and the mainland a clutter of dead, broken-off willow stems makes a safe roost where the terns congregate to rest, preen, and socialize throughout the summer in spite of there being no nesting colony within the park.

A few minutes later I startle a black-crowned night heron. It flies off heavily, with a raucous, offended *kraawk*, black, grey, and white in sequence from above to below.

I pass the willow swamp with the snipe and one of the pair is on top of an old stump. It opens its beak repeatedly, each time voicing a loud, nasal *weeeeet*. It has seen me and regards me as a potentially dangerous intruder.

Near the house two Barrow's goldeneye swim away from the bank. The drake is predominantly black on the upper body, his head purple-glossed, but most obviously he has a large white crescent between eye and beak in place of the common goldeneye's silver dollar circle of white. Almost every year at this time a pair of these diving ducks of the mountains comes to the lake for a week or ten days. Then they vanish, not to reappear.

When I come outside again the sun is setting. The wind has dropped and the few clouds have drifted away. I sit by the shore until dark. It is the end of my day. Sound, movement, colour—I observe, but do not reflect past the immediacy of the event. A female robin lands nearby, tugs free a dead grass blade, and flies away. A male red-wing hoods his wings, puffs his scarlet epaulettes as if he had been inflated from within, and enunciates his flowing song. A dozen other species sing and call; red-necked grebes clatter, the closest shatteringly loud. Far away in the cattail-choked northwest neck of the lake a bittern pumps his rhythmic *chunk-a-lunk* love song. A great blue heron flies ponderously over the point just above the tree crowns.

The light begins to dim and the frog music swells, now consisting entirely of chorus frogs. Toads trill intermittently. Coyotes yelp excitedly, fall silent, do not call again.

My muscles relax one by one; my mind empties of directed thoughts; my body feels formless, as if it were slipping and flowing away in all directions like an amoeba, leaving only a nucleus of senses in my head. Perhaps I doze, for it is almost dark when I am recalled to awareness. A penetrating *cowp . . cowp . . cowp . . . cowp . . . . cowp . . . . . cowp . . . . . cup . . up* bores in from the bay. Each time it starts rapidly, but soon unwinds, slows, stops. The source is another grebe, the pied-billed. Smaller even than a teal, it prefers cattail and bulrush beds to open water during the nesting season, and I rarely see it unless I have the time to indulge in patience.

Stars crowd the sky emphasizing its blackness, and where they end I know the land must begin, completing the sphere of night.

After five days of humid, sunny heat a cold front pushed through. It generated thunderstorms, but they came in daylight and the thunder was more spectacular and alarming than the lightning. Birds sang again once the storms passed in spite of the lingering rain. The night was heavy with its steady falling, multiplied by endless dripping from countless leaves.

It is early morning. The rain is finished, but the clouds are low and the dripping goes on. A greyness ghosts through the house. The air stirs and a sudden deluge falls from shaken branches. However, the overcast must be thinning as the light gradually brightens from above. Soon intervals of weak sunlight cast thin shadows, yet outside it is cold and no clear sky is visible.

By mid-morning the winds aloft begin pulling at the cotton wool clouds; major rifts and rents are torn and blue shows through. In a couple of hours the clouds are broken up; bright sun dominates; feet and legs soaked by the undergrowth become a minor discomfort.

Many young leaves are as shiny as if varnished. When they twist and turn in the breeze they glitter like the vanes on a metallic mobile. During the past few days plant growth has been rapid and it is now more appropriate to refer to foliage than leaves. Twigs and branches are retreating behind the expanding green, shrubs beginning to shadow the forest floor where fairy bells and western violets burst in points of white. Wherever I look white flowers gleam—gooseberry, skunk currant, sprays of pin cherry and saskatoon, columns of choke cherry.

The trail brings me to a beaver pond ringed with poplar killed when the water was raised by the dam a few years ago. Angry twittering draws my attention to a kingbird perched in one of the dead trees. It is a sophisticated flycatcher in charcoal suiting and spotless white shirt front. When it darts out at a flying insect it flashes a broad white band at the end of the tail. The kingbird's summer habitat is forest edge, but here it shows an almost exclusive preference for the meeting of forest and water rather than upland meadows or shrub slopes. A crow flaps slowly along the far side of the pond. Instantly the kingbird is after it, diving on it repeatedly and twittering frantically, although it does not actually strike the larger bird. The crow flies a little faster, but takes no other evasive action. Soon the kingbird is back, and perches again where I first saw it. It preens thoroughly before resuming its interrupted feeding.

Woodpeckers have nested previously in some of the poplars and their old holes are now being used by tree swallows. A dark-light head appears in the black circle of one entrance; then the swallow pops out and swoops low over the water. I soon lose her among the others, but a minute later a female arrives at the cavity trailing a long string of grass. She stays only long enough to add it to the nest and is out again.

A long, clearly whistled *ur-eeeeee* comes from the nearby cat-

tails. I stay still, waiting. The call is repeated several times, moving closer. Then I glimpse a sora rail, a plump little long-legged bird slipping between the broken stems and leaves. In a few minutes I see the whole bird clearly for a few seconds. It walks lightly, lifting its long-toed feet high as if reluctant to get them wet. But it keeps its head down, again and again pecking its yellow beak into the water to pick up insects and snails, like an aquatic chicken. The rail slides into an impenetrable mass of leaves, and then only its interrupted whistling traces its hidden route.

Pairs of mallards, widgeons, blue-winged teal, and green-winged teal surface feed; a single drake bufflehead floats in solitude near the centre of the pond. Red-winged blackbirds are noisy in the cattails: singing, chasing, courting. I see females carrying nesting material. A male grackle flies over, his long tail folded in a deep "v," the sun playing on his irridescent purple and bronze plumage. Dragonflies skim over the water with their shimmering wings flickering like lights.

From the forest medley of least flycatchers, yellow warblers, robins, orioles, rose-breasted grosbeaks, white-throated sparrows, a house wren, and a drumming ruffed grouse, I single out the first plaintive wood pewee. This flycatcher resembles a phoebe, but has much duskier underparts and two white wing-bars. He is high up, but even after a careful search I cannot locate him.

If I stay where I am indefinitely I will see things I have not yet observed, and come to know more intimately those that I have; but in the end will be restricted to the occupants of forest and pond. On this occasion I elect to walk on.

The trail edges past a sedge meadow with scattered willows, some of which are heavily fruited with green seed catkins. A bird trills like a slow chipping sparrow or junco, and unexpectedly he is there at the top of a willow—a swamp sparrow. A tail flick and he dives down out of sight in the sedges. Farther in an alder flycatcher wheezes, *fe-bee...fe-bee......fe-bee.* Nearer than either, a LeConte's sparrow suddenly vents his thin buzz, then is moving about in the lower, sedge-entwined stems of a willow.

Before I can see the spruce I hear a black-throated green warbler drawling his heat-blurred phrases, for the song at once conjures the sensation of a still, enervating July noon.

The white spruce are waking. On the twig tips soft, yellow-green needles are pushing off the brown parchment bud caps which come away in a single piece, patterned like the shed skin of a snake.

A small bird flies from a spruce sapling and pauses in a birch where I can see it clearly. The blackpoll warbler, like the black-and-white, is notable for design rather than colour. Several other spruce branches shake as birds fly from them, but the foliage is too dense to see through; and by the time I am among the mature trees, they have gone on, or into the upper branches.

On the ground sun spotlights ranks of woodland horsetail. Alone, each plant is a light and airy creation in filigree—whorled spokes recurving from the erect stem, dividing again, and again. Massed and interlacing they weave a continuous sheer netting. About half the stems are topped with pale brown cones where spores will form. These lurkers in today's undergrowth survive from a time when their close relatives many times overtopped them in swampy forests, and modern trees and flowers had yet to evolve.

A squirrel stutters, quickly working itself into a state of vituperation. Perhaps it is a female with young, reacting violently to every stranger.

Large black carpenter ants run up and down a spruce stump, in and out the many entrances to their galleries. When the tree blew down a year ago it tore the nest in half, although it was the ants' tunnelling through the dead heartwood that initially weakened the trunk. Since then the ants have plugged the upper tunnels with sawdust and the colony continues to thrive.

Beyond the spruce dense mixedwood spreads over a hill and alongside a small bog. Some of the spruce are mature, but most are striplings shouldering their way up through the white birch and poplar.

Sap runs down a big side limb of one old birch like colourless blood from a deep wound. The limb has been completely girdled by a porcupine, probably during the winter. Lower down, the trunk of this and several other nearby birches are circleted with neatly spaced, small, round holes. The old ones have blackened, but those cut this spring show the fresh ivory of the inner wood. True to its name, the sapsucker chisels out these wells to obtain tree sap, but mainly in early spring and later in autumn when insects are scarce. During the summer it consumes large quantities of ants, mayflies, adult beetles, moths, and dragonflies, preferring adults to larvae, and never drilling for wood-boring insects. Nestling sapsuckers are fed insects exclusively. In the park the favourite sap tree is white birch, but sometimes I find aspen or large alder ringed; and once I saw a white spruce sapling that had been tapped. A minority of trees have been used for years, a treatment that may lead to a somewhat premature death for the tree, yet few are severely damaged at any one time.

Under the birches spreads a luxurient ground cover of twinflower, bishop's cap, pink wintergreen, and bunchberry, all now unfolding newly minted, lemon-green leaves. The flower stems of bishop's cap rise wire-thin and knobbed with buds.

As I near the bog I hear what I have been expecting, a penetrating, whistled, *quick-three-cheers*. Surely other olive-sided flycatchers must summer in the park, but I have never found them. I finally see this bird on the highest snag of a dead poplar. He is a little smaller than a kingbird, plump and dark olive-brown all over except for the white throat and central breast. As I scan the tree crowns I realize how fat the aspen seed capsules have grown. In a few days they will begin to split open.

In a damp hollow dim with tangled alder I find a patch of moschatel, a low, tender plant with soft, lobed leaves and tiny greenish-yellow flowers. Soon it will be overgrown and sheltered by the robust sedges, cow parsnips, coltsfoot, and jewelweeds.

In the respite of evening, mine and the sun's, I walk along the lake. A veery sings among the Swainson's thrushes, his reeding voice unwinding down the scale even as the others ascend, introducing punctuation to the run-on days and nights.

It is nearly dark by the time I return home. In the cattails and sedges fringing the bay a pied-billed grebe calls, and added tonight is the thin, descending whinny of a sora rail serenading his mate or proclaiming tenancy.

At sunrise on this last day of May the air is spring-water cool, the calm absolute; the lake a flawless mirror to the forest, to the birds swimming and in flight, to the dragonflies tacking close to the surface. A muskrat paddles across the bay, long tail undulating sideways, its wake rippling on and on.

Half hidden, a red-necked grebe sits on its nest in a sedge clump a little away from the shoreline congestion. Without the bird the nest would be inconspicuous, as it rises only two or three inches above the water in a rough platform that could easily be mistaken for flotsam. Two weeks ago I saw the pair mating on the structure just begun; then in the days after, carrying great beakfuls of dead sedge, cattail, and pondweed they had brought up from the bottom. The nest remains more or less waterlogged; however, this does not seem to matter as it is mainly a receptacle for eggs, and the chicks leave it almost as soon as their down is dry.

The point is small enough that, no matter where I am on it, I can hear all the singing birds. Because it juts out into the lake it also seems to attract overflying birds, while those intending to cross the lake often work their way to the end through the trees before taking wing.

In a sibilant rush a flock of seven newly arrived cedar waxwings lands in the top of an aspen so that I see nothing of them but yellow-tipped dark tails, and white undertail coverts merging into subdued yellow, merging into warm fawn, merging into the black chin. They turn their heads from side to side looking at each other, and converse in thin, high trills. Then they fly out over the water, all together.

Three song sparrows sing responsively, two on the west side of the point, one on the east. A robin carols a powerful, controlled tenor. Two least flycatchers hiccup incessantly. Suddenly a red-eyed vireo begins singing. An underlying torsion seems to weave the short phrases into alternating query and answer. He continues for nearly two minutes without pause, and without changing position noticeably. Then, after a gap I hear him again, farther inland now.

For the past few minutes a song has been fretting at my memory without resolution. It is like a robin's, but hoarse; and rougher than a rose-breasted grosbeak's. When it is interrupted by a double *chet-it*, I remember—although it is not the western tanager's song I know, but the almost identical scarlet tanager's. Surprisingly, I find it difficult to pin-point the origin of the song which mingles with others, and it takes me nearly five minutes to locate the bird. Each time a pause widens I expect him to have flown, but he is just moving about slowly in the tops of the aspen. Once I sight his vermilion head and bright yellow body I wonder how he could have been so elusive, yet against the sunlit leaves contrast is weakened and the bird's vividness diluted. I watch him until my neck is stiff and sore, and return after breakfast. However, by then the tanager seems to have disappeared, and in his place a warbling vireo rolls out his rambling notes from the same high branches. I never see the tanager again.

Pioneer trails, cut by hand, neither graded nor gravelled, pounded hard only by wagon wheels and hooves, nevertheless show a stubborn persistence. I walk west along one of these old routes which has been used only intermittently by park vehicles in the last thirty years. It is still open. Grasses and wildflowers have closed over the bare, rutted earth, but poplar shoots, willows and other shrubs have been slow to encroach, their advance further retarded by browsing, for young growth, rich in proteins and other nutrients, is the most preferred. The track winds over low hills and edges ponds, bogs and fens seeking the easiest and firmest footing, perhaps initially following an old bison trail. It takes me back scarcely a generation when the ambling gait of a horse, or ox, or man had to be sufficient for business and pleasure alike.

In dense young aspen around a small bog a Connecticut warbler flaunts his song. It rings out clearly, *chup-e-ty*, *chup-e-ty*, *chup-e-ty*, *chup*, a proclamation of vitality and self-assurance. I close in to try to see the bird, yet quiet as I am, he eludes me. Ten minutes later he is still near, but has worked around me in a circle, and I have seen only fragments—a flicking tail, white-ringed eye in a grey head and somewhere, yellow. All this time he has continued singing, curious but very wary.

During the past week the vegetation has suddenly matured. First leaves have expanded to their full size on trees and shrubs; wildflowers and grasses have risen to a jungle tangle, reaching for the light above the shrubs. From now on new shoot and leaf growth will add to the density of the foliage, but its summer character is fully established. When I look up through aspen the small, round leaves on the sky are a living expression of pointillism.

White spruce have shed their bud caps. So great is the contrast between the pallid shoots and the old, bottle-green needles that, at a distance, the effect is of a painting rendered in the minute detail of super-realism. A breeze stirs and a haze of pollen sifts out from the upper branches.

At the edge of a meadow I step as through a door from a dim, cool house into a sun-flushed garden. Swallowtail, cabbage white, tailed-blue, and a few alfalfa butterflies flutter, and land, and dance again like liberated flowers. Dark-bodied dragonflies dart and hover. Clay-colored sparrows buzz unseen. A chipping sparrow trills drily. In a clump of buckbrush a savannah sparrow faintly echoes its wetland counterpart, the song sparrow. And farther off, towards the pond, LeConte's sparrows shrill in the sedges.

A dark bird moves up to the top of a young aspen. Before I can identify it visually a protracted, disjointed concoction of squeaks, whistles, and throaty chuckles mark it a catbird. He concludes his delivery with two hoarse mews, the notes that have earned the species its common name. The catbird it the most northerly of the thrasher family, which also includes the familiar mockingbird and brown thrasher. The catbird sings again, then planes down into shrubs. Here I see him now and then, moving about slowly looking for insects, and flicking his long tail. Now that the catbird is closer to my own eye-level I distinguish the coal-black crown, beak and eye, and, when his tail flips up, red-brown coverts. The rest of the plumage is a uniform dark grey.

I walk across the meadow, putting up butterflies at almost every step. Early blue violets and strawberries are still flowering profusely, although some of the corydalis are nearly finished. A few have been trampled by bison. The animals must have left only a short time ago—bent grasses have not straightened, and the sweet-pungent cattle smell lingers thickly.

Suddenly a lesser yellowlegs flies up with cries of anguish from long tufted grass in a slight hollow. Quickly it rises and begins circling, flinging spears of sound at my head. Its mate wings in from the direction of the pond and compounds the distress. Both birds fly close trying to confuse and distract me, and I back away for fear of treading on the well-hidden nest in my search for it. One of the yellowlegs perches in an aspen, teetering on the uppermost branch and still crying loudly. As I enter the forest again and move out of their sight, the sandpipers' frenzy wanes; their alarm calls falter, then finally cease.

It is almost noon and the heat today has completed the ripening of the aspen seed. Capsules split; the breeze spins a flurry of white fluff through the trees; days of summer snow will follow.

Something whirrs by my head and I instinctively duck; but it is only a ruby-throated hummingbird, its blurred wings and headlong flight more like a bumblebee than a bird. Abruptly it lands on a twig, and at rest seems much smaller than it did in flight. Now I see that it is a male. As he sits there turning his head from side to side his throat is at times black, then a slight movement fires it metallic crimson, a colour brought to life by diffraction and one no pigment can duplicate. The hummingbird leaves his perch, zooms once around me, then bores off through the trees on a tangent.

During the mid-day lull in bird song I become aware of the continuous, rain-rustle of aspen leaves. Except in complete calm it is always there, but I am conscious of it only when I deliberately listen. An arctic skipper alights on a drooping head of bluebell flowers and spreads its wings. It is a tiny butterfly, but showy, with rectangular dashes of bright orange on a background of rich brown forming a checkerboard. In addition, the hind margins of the wings are thickly fringed with orange hairs. I see several more of these skippers during the afternoon. Before I start again after lunch, a white-breasted nuthatch calls in the distance.

I push through dense shrub undergrowth to the edge of a bog. It is park-like with most of the black spruce widely spaced. In between, the dull green Labrador tea masks the hummocks and hollows so that overall the illusion is of a firm, level plain. But when I move, white cloudberry blossoms twinkle from underneath in the blanket of moss. A bird who could be a junco or a palm warbler trills from the spire of a spruce. I find it seconds before it flies, and it turns out to be the warbler.

The next depression is filled with sedges and grasses between thickets of willow and alder. Where pools lay until recently there are big clumps of marsh marigold and coltsfoot. A few fresh marigold flowers spatter deep yellow, but most have fallen. Green sawfly larvae are feeding on some of the saucer-sized leaves, where they eat all but the veins and transparent, colourless lower epidermis, creating paned windows. The sprawling coltsfoot leaves arch up from underground rhizomes. Under one, beside a rotting log I find a little cluster of flat-capped mushrooms. Straight regular grooves run out from umbo to rim like the spokes on a wheel. Colour emphasizes the likeness—sepia spokes separated by near-white. To identify the species I should pick at least one, examine the gills, and, if it is mature, collect a spore print. But the need for precision is not great enough, at least not today, and I am more than satisfied with a colour photo. The fact that the fungus' present perfection will be short-lived does not affect my decision. Even ephemeral life is entitled to live its hour. I once picked a spectacular and durable-looking bracket fungus. Its polished russet top reflected the blue sky; its minutely pored underside curved in a smooth sheet of creamy satin. But in a day or two the white began to discolour, and in spite of gentle handling, the edge to chip and break. Gradually the russet gloss diminished. In the end only my photograph remained to bring back the pristine original.

A yellowthroat warbler sings, with the cadence of the Connecticut warbler, but none of his panache. I exhale "swish" several times and the unfamiliar sound attracts the yellowthroat. He stops singing and approaches uttering short *tchu* notes. Then twigs shudder only a few feet away, and he is bending forward to peer at me intently, black eyes hidden in a black bandit's mask. For the rest he is plain yellow below and olive above. I continue to swish and squeak, and he moves around me through the willows, once coming almost near enough to touch. But soon his interest changes to other matters and he moves off again. I am no longer a novelty.

More forest flowers have opened: dewberries, low-bush cranberries, baneberries, and woodland coltsfoot. Where sun strikes at the edge of the track a scattering of bunchberries, and the spherical heads of sarsaparillas under the umbrellas of their horizontal leaves, dapple the foliage with white. At present the only coloured woodland flowers are bluebells, together with a few late red currants, and I ponder the significance of white. Biologists are still far from a complete understanding of the intricate relationships between insects and flowers, even though knowledge has been accumulating for well over a century. Some insects which visit flowers have little or no involvement in cross-pollination, coming only to lap nectar, to eat pollen or the blossom itself, to prey on other visitors, or further, to warm up in those flowers that trap heated air. These insects may land on other flowers of the same species later, yet cannot fertilize them since they carry no pollen.

Nevertheless, a variety of insects in addition to the well-known bees, moths and butterflies do contribute to pollination. To attract these assistants the flowers advertise by one or more of the following: colour (although the insect spectrum is not identical to ours, extending into the ultra-violet, yet not so far into the red), scent, size, shape, and contrast with surrounding foliage. A number of insect-flower interdependencies have been elucidated, and perhaps the most singular in North America is that between the yucca and the yucca moth. Neither can reproduce without the other. The adult female is the only and essential pollinator of the plant's flowers, while the moth caterpillars feed exclusively on the developing seeds, eating most, but not all. However, many plants and insects are much less exacting, although legumes do require specific pollinators. But why are so many spring flowers, at least here, white? In general, yellow seems most attractive to insects and white least. On the other hand, flies come more readily to white flowers than many other insects, and they, especially midges, mosquitoes, and hover flies, outnumber other pollinators early in the season, although moths are also common. Within the forest, white offers the maximum contast and the petals are clearly outlined against the clutter of dark foliage. Additionally, some white flowers possess ultra-violet lines or spots which are attractive to insects. Contrast then may be one of the main attractants that forest flowers possess, although some are also delicately scented.

I turn onto a narrow bison trail which leads to the shore of a small lake where the banks slope steeply to the water. Part of my way passes through shrubby growth, hedged to about four feet by browsing, and for the same reason stimulated to sprout profusely into near impenetrability. A mourning warbler follows me invisibly, but his frequent singing places him precisely. Although this species is closely related to the Connecticut, the song is quite different, loud yet lower pitched, shorter and less jerky.

Near the water I stand under the shade of a solitary aspen. Having grown in the open the tree spreads its branches widely and its crown is deep. Low down its trunk has been polished by rubbing bison, and its roots exposed by their trampling. Across the lake the land is level and scattered old white spruce spike up through the dense wall of green. They also dominate a tiny island linked to the mainland by a swath of bulrushes and sedges.

A Canada goose glides slowly away from the cover of the rushes like a ship under sail. I search minutely, and in a few minutes discover the almost hidden heads of two more geese. I stay for half an hour, but these geese never move; however, a second joins the one now feeding in the shallows. Presumably the birds are nesting. There are ducks too on the water: mallards, widgeons, blue-winged teal, green-winged teal, a couple of gadwall pairs, a few pintails, lesser scaups, goldeneyes, and buffleheads. Red-necked grebes bray, and a horned grebe out towards the centre dives repeatedly. Each time it surfaces it shakes its head and elevates its buffy ear crests. Red-winged blackbirds sing from the cattails and willows, and several grackles wade into the water to dip for insects and snails. A spotted sandpiper lands near them and begins feeding too, stepping quickly, tail wagging up and down constantly, its reflection rising to meet it each time it plunges its beak into the water.

A blue jay jolts the calm with strident jeers, but I do not see it. A group of cowbirds flies into my aspen, their thin voices filtering down the tower of leaves; then they traverse the lake, a knot of dark specks undulating independently yet keeping flock unity. A minute or so later three cedar waxwings fly over without pausing.

I step from the pool of shade into the sun's glare and heat, and start back. The breeze is too weak to fan my skin, and even slow walking brings perspiration. Not yet acclimatized to summer, I seek every scrap of shade.

In the meadow by the pond three huge bull bison lie in wallows,

but I stop in the trees before they see me. One stretches on his side and rolls. Legs flail the air, then he flops down on his side again. Soon he is almost lost in a cloud of dust; half rises to turn and dust his other side, for his shoulder hump makes a complete roll impossible. Finished, the bull stands up and shakes. Dust balloons out and settles to the ground as the animal continues to stand. The other two have been watching, like me, but never cease their cud chewing. I detour widely, to avoid both the bison and the yellowlegs' nest.

I sit for a long time at the end of the point in the early evening in spite of the mosquitoes clouding the windless air. The lowering sun blanches the sky almost white; the lake in repose is scarcely darker. Slowly sinking, the sun swells to a glowing disk and disappears behind the trees. After its departure the last flare of brilliance dies and blood-orange bleeds out above the forest in a spreading stain that melds with the zenith blue. The water repeats the sky, doubling the sun's aftermath. A long half hour later the land visibly darkens as the pendulum reverses, the glow congeals and the night arches moonless from the east.

Like a great, slender-winged bat a nighthawk jinks across the empty void, high, scooping up insects. It utters single, nasal notes that seem independent of the bird. Suddenly it dives headlong, a smooth descent ending in a steep curve and a roar of air through braking wings. Then it climbs again jerkily, its scimitar wings beating deeply as if the bird needed all their strength, yet it is not without grace. It moves farther and farther out over the lake into the night, but I cannot recall the exact moment when I could see and hear it no longer. On this last day of May the high tide of bird migration has reached its crest. The nighthawk is the last arrival.

June

I walk along the lakeshore after breakfast. Ring-billed and Franklin's gulls, together with common terns are far-out white motes curving against the spruce. Ducks and red-necked grebes speck the still water.

I bend and peer into the shallows where the dense sapling growth of pondweed foreshadows the lush forest that in a month will reach the surface. Aquatic invertebrates concentrate around and through the pondweed. At the edge, only a few feet away, a male stickleback fusses at a globular mass supported by several stems. It is made from bits of plant cemented together with the fish's own secretions, and is the nest in which he will entice females to lay their eggs, perhaps has already. The females leave once they have shed their eggs, but the male guards and aerates them until they hatch in late June, and then for several days shepherds the transparent fry. Other sticklebacks hang suspended, held by fanning pectoral fins; or cruise slowly; or shoot forward suddenly to snap up water fleas and small insects.

A few sticklebacks float almost at the surface, lethargic, abdomens swollen and white. They are hosting large tapeworms whose life history is as convoluted as the worm itself. The tapeworm eggs, floating free in the water, develop into microscopic larvae when swallowed by certain copepods, themselves almost microscopic crustaceans. When a stickleback eats an infected copepod digestion releases the larvae which then enter the body cavity of the fish. One, or occasionally two, may survive to grow, over several weeks, to near maturity. Their weight alone eventually slows the fish, while parasitized females produce few viable eggs. I suspect that parasitized fish are also less resistant to environmental stresses. However, only when a stickleback is eaten by a bird does the tapeworm finally mature, and in a mere day and a half as it passes through the bird's digestive tract. There, unless taken in large quantities over a very short period, the parasite causes little discomfort to its final host, although that host is crucial to the species' perpetuation. I have not been able to find published

studies to indicate whether the tapeworm will also complete development in a mammal predator such as a mink.

As I rise I notice, on the underside of an arching grass blade, an adult dragonfly emerging from its nymphal skin which has split along the top of the thorax. The dragonfly hangs head down, attached only at the tip of the abdomen, a pale effigy of the scintillating aerial predator it will become in an hour or so. It is motionless, legs tightly folded, shrunken wings crumpled, abdomen wrinkled like a closed accordion. For fifteen minutes it does not move. Then its legs slowly open, stretch, half close again, open. The body jerks spasmodically until suddenly it flips up and the legs fasten onto the head of the nymphal case. All movement ceases for several minutes. Then a few shudders release the abdomen, and the adult is at last free of its juvenile trappings. But it continues to cling to the empty husk. During the next half hour blood pumped along the wing veins expands them and spreads the transparent membrane as a flower opens visibly with time-lapse photography. Finally the wings are fully extended, the abdomen rounded to a smooth, segmented cylinder, and colour deepening to a dominant blackish-brown, although there will be flecks of a brighter hue, probably blue or green. The wings vibrate, rustling, but the insect is not yet ready for flight. It crawls slowly off its discarded skeleton and a little way up the grass blade; then rests again. I look along the bank and find several other dragonflies of the same species also emerging. Some have only just begun; others are in various stages of completion. One, like a machine off the end of an assembly line, suddenly flies away, adept without practice. I straighten up, standing. From above none of the dragonflies are visible. Is their choice of emergence site intentional or fortuitous? During the interval between leaving the water and their first flight they are completely helpless.

I am still standing as a mallard swims around a corner close to the shore, leading a flotilla of eleven fluffy ducklings no more than a day or two old. They peep softly and continuously as they twist and turn dipping their stubby beaks into the water. I am partly screened by saplings and the mother does not see me, although she frequently raises her head from surface feeding to look about for danger to her brood. They swim on out of sight, keeping only a few feet from shore.

Many young willow, poplar, and dogwood shoot tips are thick with aphids, wingless and pale green. Some colonies are tended by ants which intermittently run up and over the massed aphids to touch them lightly with antennae. Perhaps they have already "milked" them and are now on the lookout for predators and parasites. At unattended colonies I find a few newly hatched ladybird larvae, one to a colony, but no lacewings. Several days ago, however, I saw some lacewing eggs, tiny white beads on the ends of thread stalks spaced around the margin of a buffaloberry leaf.

A song sparrow flies into a willow and begins its *chuck*-ing alarm, tail and wings flicking in agitation. Its beak bristles with small insects for its young. I back away, carefully watching where I put my feet since this sparrow is often a ground nester.

By mid-morning the dew has dried, and the aspen seeds begin to drift. Dispersal is almost at its peak; the drift is a snowstorm. Each tiny seed is surrounded by a gossamer network of white silk filaments. This fluff catches in spider webs, piles lightly on big leaves, tangles in the needled spruce turning them hoary, and lies in windrows on the ground. Under a shedding tree I feel suffocated. How many million seeds does one tree release? How many billion billions a forest? Yet only a few will grow to trees. The seed must germinate on mineral soil, for it has almost no food reserves; and germination must occur quickly or the embryo loses its vitality. It seems like waste on a vast, incomprehensible scale because all poplars regenerate easily and profusely from root suckers. Yet even a few trees from seed bring genetic diversity to the population, and situations occur where such production comes into its own—blanketing the charred ground after a forest fire that has consumed hundreds of square miles; filling, in the past, buffalo wallows to initiate new groves on the prairies where trees are often killed by fire, drought, and severe winters.

Even in the forest not all the fluff is useless. In a mile of walking I see several birds fly down to the trail and gather beakfuls for nesting material—a purple finch, least flycatchers, yellow warblers, and a wood pewee. Often so much fluff goes into the structure of a nest, particularly a yellow warbler's or least flycatcher's, that it appears made of cotton wool.

A cock ruffed grouse thumps out his message, perhaps more to himself now as most females will be incubating in solitude.

The water of a big beaver pond glints through the trees ahead. Long before I reach it a large bird, dark under the leafy canopy, flies from its perch and weaves through the columned trunks. Its silent departure indicates an owl, but my glimpse of chestnut-barred underparts and sharply black and white banded tail define a broad-winged hawk. I search the tree crowns, but find no trace of a nest.

Halfway down the slope to the edge of the pond another bird rockets up from almost beside me, and not until it is well out over the water do I recognize the solitary sandpiper. This species does not make its own nest, but uses one that has survived the winter, often a robin's sturdy grass and mud structure. I suppose the young climb out and drop to the ground as do those of hole-nesting ducks, for once their down is dry, like all shorebirds, they are ready to run about and feed themselves. Then their parents lead them to the pond or lake that is always nearby.

I sit on a beaver-felled tree and look out over the pond. The usual ducks are present: mallards, widgeons, both teals, a few pairs of scaups, and a bufflehead. Except for the paired scaups most are drakes. A dozen common goldeneyes, all apparently hens, seem too many for the potential nesting trees, and I conclude that at least some must be yearling males as this species requires two years to reach adulthood. Further scrutiny reveals a ring-necked duck, two redheads, and a shoveler, all drakes. A pair of red-necked grebes float by their nest anchored to a drowned willow.

Duckweed has spread in a yellow-green carpet far out from the cattails where a multitude of new leaves is beginning to cloak last year's bleached and beaten rags. I raise my binoculars and scan the forest behind; stop at the framed head and neck of a buck white-tailed deer. What I can see of him stands out clearly as if he were part of a tapestry. His antlers are half grown, but blunted by the velvet nourishing them, and almost the same reddish-tan as his hair. He looks in my direction, yet does not seem to see me. Frequently he shakes his head, probably in irritation from the deerflies and mosquitoes reveling in the thin, blood-rich skin covering his antlers. Once he drops his head and rubs his nose against a foreleg. In a few minutes he turns back into the forest, and my last view is of a flicking, white-bordered tail.

Today must mark a general emergence of dragonflies, as the numbers flying are suddenly legion. Most are the darners I studied early this morning. Mingled with them are a few small skimmers, deep claret of body, which were the earliest adults.

In a shallow depression in the forest where tree bases and decaying logs are thick with moss and the leaf litter almost always damp, many wood frogs hop away from every step. They range in size from just over an inch to slightly more than two inches, but the largest are rare. Probably all these frogs have bred this spring, most perhaps for the first time. The great size variation reflects differences in transformation dates last summer, and also possibly, subsequent nutrition. The large individuals are the most colourfully marked—plain earth brown on the head, back, and upper legs lightens to vermilion-tinged fawn on the sides, to creamy white on the underparts. Chin, sides, and legs are spotted and barred with black; the mask is sepia. The pattern is completed by a slender white line bisecting the full length of the head and back. The smaller wood frogs are a slightly paler brown above, with little spotting, and many lack the mid-dorsal line, although this last

feature is not related to age or sex.

With my attention on the ground I also distinguish a big squat toad beside the path. It sits among the leaves heavily, arms akimbo, mouth set in a perpetually dour line, gold-flecked eyes expressionless. Its mottled, warted skin shows a greenish undertone, probably in response to the surrounding foliage, for I remember early May toads being predominantly brown and grey.

Although forest insects are becoming more diverse and numerous, flies still seem to be in the majority. Today, swarms of small grey midges swell their ranks. Many end tangled in an orb spider's geometrical trap. Green lacewings and tiny bark-coloured moths flutter off as I brush past shrubs. Ladybird beetles frequent leaves, flowers and twigs. I see other small beetles on trunks, leaves, and at flowers; various true bugs, green and brown; wasps, bees, ichneumon flies, and sawflies. Sun strikes hotly into a gap where fallen poplars have not yet been replaced. Nearly every sunny shrub leaf is crowded with vivid blue and black damselflies. Perhaps they are just basking, for when disturbed they fly only briefly before alighting again.

By early afternoon tiger swallowtails have gathered at the edge of the forest cresting a shrubby slope. They cling to choke cherry, honeysuckle, and saskatoon flowers, probing for nectar. And they climb in spiralling flights as pursued and pursuer trace out the same course. Other butterflies have also been attracted here: tailed-blues, alfalfas, cabbage whites, tortoiseshells, fritillaries, and a species I have not seen this year until today, a common alpine, sooty brown with a flash of red-orange on each forewing. However, all these diminish in the presence of the swallowtails.

The flowering shrubs have drawn additional insects. When I come close I find hover flies, bumblebees, solitary bees, and many small beetles and bugs. Once a hummingbird moth darts in to a choke cherry. Its proboscis uncurls and extends beak-like into a flower. The insect hangs on invisible wings, but unlike the bird, its two front legs reach forward to steady itself. It spends over a minute at one spike of flowers, methodically testing each blossom; then leaves, its flight fast and direct like that of a real hummingbird.

At the bottom of the slope young poplar forest closes around me, but soon opens again, encircling a meadow of ponded sunshine. As I stand looking into the light I hear laboured grunting not far away. It comes intermittently and irregularly, and puzzles me for some time. Eventually I decide it must be a bison, moose or elk giving birth. I feel it prudent to keep my distance, and move away as quietly as possible.

Early evening brings a resurgence of bird song, frog chorus, and toad trilling. Sora rails and a pied-billed grebe call from the cattails; a snipe thrums overhead; gulls keen out over the lake. Coyotes herald the sunset, their soprano yelping muffled and mellowed now by the clothed forest between us.

By mid-day ostrich plumes of cirrus fanned out of the west. Two hours later high white stratus sheets dimmed the sun, while scattered, sullen cumulus slid under them. But the breeze still wafted in from the southwest, hot and humid. Not until late afternoon did dark storm clouds loom above the trees. Still miles away, they united in a solid front; sailing nearer they reared up and over like the crest of a beaching wave. Squalls separated from the main bulk of the cloud; the wind veered and gusted; rain slatted down, big, heavy drops. It lasted for twenty minutes then eased, but did not stop. Soon another squall drenched down, driven by the wind, and was followed by roll after roll of grey fleece, tattered, shredding into rain. For a while I stood under the dry tent of a spruce looking out, watching the drops gather on leaves, merge, run down to the tip bending under their weight. There they gathered again in fluid crystal until too large, they dropped onto another leaf, and often another, and another, in stepwise descent to the ground; thence trickled down through the dead leaves, decaying mulch and into the mineral soil. Down, always down, until so absorbed and adsorbed by humus particle, and clay particle, and boulder face they were too diffuse to respond to gravity.

Night came early and the hours of tossing, dripping darkness seemed forever.

This morning crept in, the spreading light revealing restless trees, bowing cattails and bulrushes, grebes hunched on their nests like lumps of dead wood. A robin or two, a song sparrow, a yellow warbler, and a few red-wings sang intermittently, their voices at half mast.

Part way through the afternoon the rain lessens, but it is three more hours before it finally drizzles to a stop. The low cloud scuds on and the ceiling lifts to tightly packed, clean-edged overcast. The wind, now northwest, weakens to a light breeze. But it is cold, down to the mid-forties.

With the end of the rain birdsong revives; the red-necked grebes leave their nests to feed; gulls and terns take to the air again; tree swallows ground swell over the lake and flutter around the spruce where insects have gathered in refuge.

I hear a beaver chewing. Soon a ten foot aspen begins to sway. The chewing stops; the aspen leans, swoops down into the water. The beaver splashes after it. I creep closer. The beaver is sitting in the shallows stuffing leaves into its mouth as fast as it can grind and swallow them.

The land is sodden. Undergrowth continues to rain. Wildflowers bend over, their blossoms drooping and leaves turned to pale undersides. I can hear water entering the earth as if it were being pulled right to the core by a deep-breathing, subterranean being.

My watch tells me sunset, for the cloud mask does not crack before clammy night squelches down.

Sunrise diffuses through the ground-glass screen of fog. Briefly the cut-out images of the islands, the oil-smooth water, the solidity of the fog itself are all permeated with a golden radiance. Then cloud covers the sun and the world goes grey, and dank, and cold once more. The fog persists, its density mapped out by receding landscape planes. The bank at my feet is clear-cut; near headlands appear miles off; islands half way across the lake are little more than thickenings of fog; the far shore is invisible. Waterfowl close to shore are impeccable, but at a hundred yards their colours are washed out; a little farther and they are grey silhouettes; beyond that they are loosed voices crying in some primal miasma. The lake suddenly seems huge, stretching to the infinity of the ocean.

By late morning the overcast breaks and the sun slowly thins the fog. A breeze stirs and the last shreds wisp away. The sun between the scattering, puffy clouds is bright and warm. Colours are saturated and intense; shadows precise and dark. Water drops prism the light with sparkle. The world is new made. Birds rejoice in song.

I spend the afternoon outdoors. The sun simmers a lush fragrance blended of wet, fermenting leaf mould, flower perfume, green leaves, and tree bark. I find a mature morel, its still wet skin waxy to the touch, culminating the fruition of decay.

Fairy bells, marsh marigolds, baneberries, gooseberries, and currants are almost finished with flowering everywhere. Gone is the white tracery of cherries and saskatoons. Nevertheless, a few plants that blossomed early, like strawberries and white violets, continue to flower. Others are nearing peak profusion: star-flowered Solomon's seal, twinberries, sarsaparillas, bluebells, smooth black currants, and the wind-pollinated, petalless meadowrues. Beneath the shrubs and taller herbs the first white spikes of mayflower are exclamation marks; dewberries are five-pointed stars; bunchberries recurve green sepals that in a day or two will bleach white; sandworts shine, small and delicate; both pink and white wintergreens raise budded stems from rosettes of glossy leaves. To appreciate the unique bishop's cap you must get down to the level of their eight inch stems arising from thick clusters of scalloped leaves. Spaced widely, each flower arches away from its slender support, alone and precise against the darkness of shadow and apparently anchoring a miniature spider's web, an illusion spun by the finely-divided petals.

White peavine have tendriled their way up to waist height and put out one-sided spikes of showy, keeled flowers. At the same level and above, a scattering of wild roses are open, while thousands upon thousands of the tapered, furled buds will spread in the next few days. Here and there, the tiny flowers of low-bush cranberries crowd in flat-topped discs. Even now with butterflies and bees in abundance, most forest flowers are white.

Brooklime sprawls between the shorn marsh marigolds and dense ranks of jewelweed seedlings growing along the winding seep below a beaver dam. Singly the diminutive, pale blue flowers would be unnoticed, but grouped into slender racemes they catch the eye without losing their character. Larger than a human hand, cow parsnip leaves rear over the brooklime. The trunk-like flower stems are nearly three feet high and are tipped with the budding heads. Layered above the parsnip, but below the shading alder and willow, wine-stemmed dogwoods display their convex white flower clusters, each in a stylized setting of strongly veined leaves.

At the edge of a sedge meadow marsh reed grass grows as luxuriantly as hay. Suddenly something small leaps, and plumps down a few feet away. I advance very slowly, then lean over. A mouse crouches palpitating, deep in the grass yet clearly visible. It is about the size of a deermouse, but the ears are smaller. Black along the middle of the back shades into bright tan on the sides. From what I can see the feet are white, and also the lower side of the slender tail that is one and half times as long as the body. This is my first sight of a jumping mouse, but even before I saw it in detail I had recognized it from its prodigious leap. As long as I stand over it the mouse will not risk moving. When I step backward it scurries away.

I hear crows cawing as I have done every day since their return. Theirs is one of those background sounds that becomes so commonplace I would notice only its absence. Sometimes when I am in the open or by a lake I see the birds themselves, most often a single or pair, although occasionally still, a straggling flock of up to a dozen. They nest in the park, but I do not know precisely where. I have never happened on a location, nor spent time looking for one, merely accepting the presence of crows as part of the general landscape.

The trail winds on through dense, mature poplar forest. A sapsucker flies past and as I follow it with binoculars, my line of sight is checked by a least flycatcher's nest. About fifteen feet above the ground, it is cupped in the crotch of two branches on a dead aspen sapling, completely without concealment, although the nest itself resembles frayed bark. It was the unusual shape of the filled angle that caught my attention. One of the birds is on the nest, and I can just see the round of its head as well as its long tail projecting out like a dead twig.

A female robin appears almost at my side on a low branch, calling harshly in alarm, yet never dropping the moth gripped in her beak. She continues to fuss around me as I stand, and I begin searching the shrubs. Soon I see a fledgling robin close to the ground, clinging to a hazel stem and looking about vacantly, eyes large and round. Its beak is still flared at the base; tufts of down show through its feathers; its tail is scarcely an inch long, and its body is dumpy. Probably it is not yet able to fly. As my eyes adjust to the dimness I see a second and a third fledgling near the first one. Then the male robin arrives and together the adults make an overwhelming protest. I hurry on. The male follows me and I am some way from the family before the commotion dies down.

Caught up in the robins' distress, I fail to see a pileated woodpecker until it flies up in alarm from a pulpy log it was excavating. Long shards of soft wood lie on the ground where they were flung, and the beginnings of a gaping hole reveal tunnels through to the log's interior which could have been made by beetle larvae or carpenter ants, or both, but the woodpecker did not have time to find out.

I feel I am precipitating a chain of disquiet that will roll forever before me like a bow wave, and I stop for several minutes to let the forest regain its own rhythm. The flow reverses—from going away to coming toward, passing, going around. I am witness to only segments of activity while animals are within sight and hearing, but I know those segments are not affected by my being. Then a deerfly alights on my sleeve. Her delta wings are dark-banded, abdomen striped in tan and black; but eyes are her most singular feature—very large, hemispherical and crossed with bright green and purple bars further enhanced by irridescence. Like most other female biting flies she seeks a meal of blood whose proteins are necessary for proper egg development. I flick her away, but soon she is back probing for my skin.

A yellow warbler drops down to the trail and begins picking up aspen fluff. When she has stuffed her beak full she flies off through the shrubs.

A queen bumblebee drones past, her work tripled many times over now that she has young. Until the first workers mature the queen must look after the entire nest, build new cells, lay eggs, and feed her growing offspring.

Countless tiny, pale green midges drift around and among the shrubs. They are probably swarming males waiting for possible mates to be attracted by the specific hum of their vibrating wings.

Now that I am sitting, eye level has dropped to just over two feet and I find myself looking at the undersides of hazel and rose leaves. Where the sun strikes them they are a vivid, translucent chartreuse. Veins stand out darkly like a wire framework. A swallowtail flutters over, it too gaining brilliance from the light shining through its wings.

Beside me in a patch of bluebells two pericopid moth caterpillars hang motionless, head down, on separate stems. They are either preparing to moult, or nearly ready to pupate. Their black ground colour is interrupted by patches of bright yellow down the back and along the sides; blue rings the segment joints; tufted black hairs bristle the entire body. The two convex eyes occupying most of the head are orange-brown.

Grass grows up between the bluebells. Almost on the ground several leaves have been drawn and held together with a filmy cobweb of silk threads. Within this a ball of just-hatched orb spiders are like tawny, round seeds.

A full-grown forest tent caterpillar undulates at a brisk pace up an aspen trunk. Like many moth and butterfly larvae this one is colourful and distinctively patterned: white exclamations framed in black punctuate each segment mid-dorsally; the pale cobalt-blue ground colour is adorned on the sides with pencil lines of black, chestnut, and cream; sparse tufts of long black hairs crest the back, while shorter, but denser, fawn hairs skirt out over the legs; the eyes are dull blue-grey. At intervals of a decade, more or less, the population swells to spectacular numbers that defoliate large tracts of forest and cover the ground as the caterpillars search for protected pupation sites. More usually parasites, diseases, unfavourable weather, and birds—here mainly orioles, rose-breasted grosbeaks, red-eyed vireos, yellow warblers, and black-

and-white warblers—keep the numbers to inconspicuous levels. I look around, but can see no other tent caterpillars.

I come out into sun on the side of a shrubby slope. My sudden appearance must have startled the sharp-shinned hawk waiting in the trees. From nothing there is heedless, headlong flight, frantically beating wings, vanishing image between the far trees. The hawk's fleeing silences all song and I find consolation of sorts in knowing that animals other than man are cursed with evoking the blight of fear. I remain still, becoming a part of the landscape as the hawk did before me and soon the small birds sing again. Three male clay-colored sparrows occupy this slope. Two have the usual three-buzz song, but the third consistently continues for six, and a few times for nine.

A rapid trilling *ki,ki,ki,ki,ki,ki,ki,ki,ki,* all on the same pitch comes from a kestrel hovering about twenty feet up and facing into the breeze. Her head markings are similar to the male's, although duller, but her back, wings, and tail are a uniform black-barred chestnut, while the same brown streaks her white underparts. She hovers for several seconds, shifts downslope, hovers again. Then she is no longer in place, but a blur dropping groundwards reaching out with extended, grappling claws; beating up again out of the grass with a dark lump in one foot. She flies to an old dead balsam poplar at the base of the hill, lands on a large branch and begins to eat the dead vole. She raises her head each time to swallow, and looks about switching her head from side to side as well as tilting it to see better overhead. The poplar must have matured in the open as all the trees now flanking it are much younger and its branches reach out like extended arms. The trunk is short and stout. About half-way up is the black circle of an old pileated woodpecker nest hole that seems to have been appropriated by the kestrels, for now I find the male, motionless as a carving, at the top of the highest branch, dark against the blue sky. The female finishes the vole, strops her beak on the branch a few times, then begins a meticulous preening.

Hundreds of enamel blue and black damselflies course back and forth in the sun. From time to time some alight on a grass blade, shrub leaf or twig, swivel their wings over the abdomen

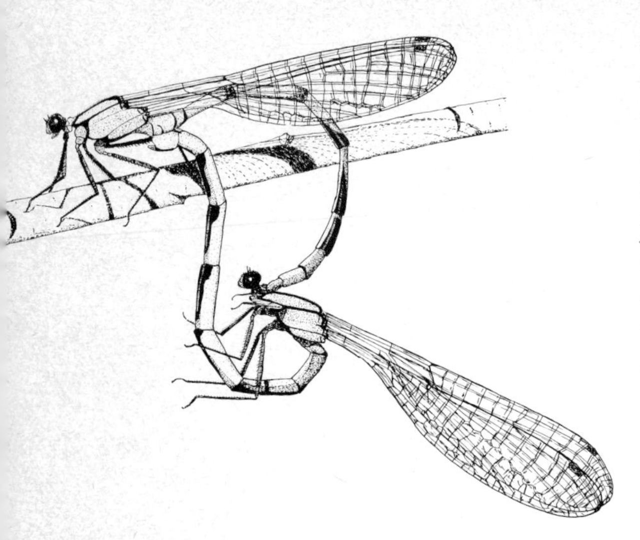

and rest. A few pairs fly in tandem, and one of these lands nearby to complete mating. The final position is bizarre and seems incongruous until explained. Claspers at the tip of the male's abdomen grasp the female behind the head. She arcs her abdomen forward and affixes its enlarged end to the underside of the male just behind the thorax. The reproductive loop is joined, and sperm is transferred to the female. How? The male's sperm ducts open conventionally at the end of his abdomen, but prior to mating he bends it forward and empties sperm into a special internal sack which will be covered by the female's genital opening. And why such elaboration of a simple act? Some experts believe that dragonflies and damselflies evolved from insects whose males deposited sperm in a drop of fluid that the female picked up, as in the contemporary wingless bristletails. To mate in flight the male would have to place the sperm droplet on himself, and the most easily reached place is the front of the abdomen. Given time—and there was more than enough since the Odonata is an extremely ancient group—the internal sperm sack developed together with the contorted mating posture I witness now. The damselfly pair remain still for a minute or two, and will probably stay united in tandem during egg laying in a nearby pond.

On a background of deep shadow, sunshine highlights a suspension of hover flies. For long moments they hang almost motionless on hazed wings, each separated from its neighbours by several inches. Abruptly one darts sideways. Immediately adjacent flies adjust to keep their former spacing, and their neighbours, and theirs, until the entire congregation has shifted position. Before this is accomplished the first movers are still again. This gathering is probably a swarm of males awaiting the arrival of females.

The overgrown trail resumes at the bottom of the slope. I step out into the open angling across and down towards it. The male kestrel flies and circles over my head crying. His mate is not in sight now. Butterflies start up. Just at the edge of forest again a pair of tailed-blues is mating on a grass stem. They make a balanced composition—facing away from each other, the male's wings half spread, revealing the subtle blue shading of the upper surface; the female's tightly shut, presenting the downy-white undersides with markings of soot specks and orange crescents.

The poplar seed storm blows once more and today is augmented by the fluffed seeds of early-flowering willows, although their contribution is scarcely noticeable. The spruce which had been washed clean by the rain will soon be smothered for a second time.

On the way home I visit a small knob island near the south shore of Astotin Lake, and on this day for a single purpose. On the sun-struck southwest bank flourishes one of only a few thickets of silverberry, shrubs with hoary, blue-green, leathery leaves and now, inconspicuous clusters of yellowish-green flowers. It is not the flowers themselves that are the attractant, but their scent. I first pick it up many yards away where it has flowed downwind. As I near the plants it strengthens. Like the flamboyant swallowtail butterfly, the silverberry with its almost too-sweet, cloying fragrance echoes the tropics. And the family to which it belongs is also primarily sub-tropical. Yet, in suitable habitat, the silverberry can be found north to the treeline where its heavy scent is even more incongruous.

Toads trill above the uneven backdrop of chorus frogs as the setting sun closes off the day. Ducks are silhouetted on the bright water. Two female mallards head newly hatched broods which diverge, concentrate, elongate, and congeal as the ducklings feed in the narrow orbit of their mothers' shield of security. Slim-necked, high-headed grebes glide smoothly. Gulls fly in for the night. Common terns beat back and forth, sinking to the water, rising above the trees, their skirling cries reaching me clearly.

Then a great hemisphere of water, forest, and sky is commanded so completely by one sound that temporarily no other exists. Wherever there are ears within a mile radius they are flooded, nearly drowned by the pure ululations of two loons. A low moan quickly transcends an octave and rises even higher before suddenly dropping at the end. Then the birds seem to become excited and for two minutes yodel without pause, sequence overlapping sequence with no beginning nor end, no definite alternation. Suddenly, as so often with animal expressions, the birds quiet.

I wait for the loons to call again; they do not. Perhaps they are now fishing or have flown back to their home lake. In the wake of their exultation the grebes are suddenly insignificant, the yipping coyotes insubstantial shadows of sound.

❧ ❧ ❧ ❧ ❧ ❧ ❧

It is four days before the solstice, before the longest day of the year, before the pivot of the summer. Although I am now used to the heat I still prefer the morning hours with their rich colours, drying dew, and animal vitality.

Can winter possibly be as I remember it? A cold and dark desert here on this same ground—snow streaming in the lacerating wind, drifting like white sand; grass and flowers sepulchred, dead to their frozen roots; stripped trees and shrubs thin frames of twisted wire; chickadees puffed balls of fluff; bison, moose, and elk cobby with insulating fleece. But the recollections I summon seem too extreme and remote to threaten the present.

The realities of the moment are lush sedge meadows and cattail beds; thick duckweed carpets some of which seal over entire ponds; plankton blooms greening the lake; jungle undergrowth where plants intertwine, lean on and over one another in an apparent incoherence in which I must search out the leaves belonging with each flower; duck and grebe broods multiplying daily; and the confusion of insects dominating every element.

This morning is fresh, although the sun beating out of a cloudless sky promises heat before midday. At the edge of a shallow pond I step carefully past white-cowled water arums saintly in their cloisters of glossy, dark, orbicular leaves. A yard or so farther and water stands between chafing clumps of sedge. Hundreds of wriggling black tadpoles clog its transparency. About half show tiny hindlegs where the slender tail joins the bulbous body that looks to be all head. A smaller number, larger and greenish, are wood frog tadpoles, whereas the dark ones are chorus frogs and possibly toads. A few adult chorus frogs sing on the far side of the pond.

In the mature poplar forest the understory resembles a planted rose garden. Wherever I look the green is studded with the showy, three-inch blossoms. In bud and when first open the petals are royal magenta, but they quickly lighten to a pure, bright pink, and just before dropping may fade almost white. But now, nearing the peak of their season, there are few old flowers and as many in bud as open. Wild roses are common in this region, but their prevalence within the park may well be a result of prolonged, intensive browsing which has stunted many other more palatable shrubs.

I look into one young rose expecting to be rewarded with the disc of golden stamens like a field of ripe grain; but instead it is obliterated by a mass of dark, bristly muscid flies crawling over one another and pushing in their eagerness to lap pollen and nectar. For some reason the sight is immediately revolting. The flies are not damaging the flower; a few of them will even carry a little pollen to another rose. Nor are they ugly individually. It must be that massed the flies remind me of the same insects swarming over a dead animal. But even explained to my satisfaction, nothing changes my intial "beauty and the beast" aversion. It is some time before I look into a rose again.

Yet there are other forest flowers almost as showy as the rose: the twining honeysuckles' fountain-head clusters of orange trumpets, the purple peavines' and vetchs' short-staffed crooks, the high-bush cranberries' white saucers ringed with sterile, oversized corollas to attract pollinating insects, and the cow parsnips' convex white lenses which soar over my head where the plants root in damp hollows. Where the shrubs are less dense Canada anemones have come up in thick stands and now spread virginal-white sepals to the sun. I bend close to one. Half-shielded by the clustered stamens, a white crab spider crouches with its fangs sunk into the head of a hover fly almost as large as itself, but immobilized by venom.

Snakeroot and yellow avens would be lost in the general green were their leafy stems not so robust. The former produce marble-sized spheres of petalless flowers, the latter single small yellow blooms on long diverging pedicels.

From the second half of June and through to mid-July forest flowers are at their greatest diversity and abundance. Each species has its own particular flowering period, although this can be lengthened or shortened, advanced or retarded to some extent by the plants' siting and the current year's weather. Beyond these variables flowering duration is governed largely by whether a succession of buds is produced over weeks, or all buds open at more or less the same time. For instance, bluebells began flowering in late May and will continue into early July. Each plant produces several flower clusters, not all of which mature simultaneously. Furthermore, within each cluster there is a continuity of pink buds becoming blue flowers withering to white parchment husks and falling. Many other species follow the same pattern. On the other hand a few, like the single small head of bunchberry and the solitary terminal flower of Canada anemone individually rarely last more than two weeks.

In the cave-like shade under the shrubs and taller herbs where the air is humid and still, mosses anklet woody stems and tree trunks, and cushion deadwood. Occasional mushrooms dome through the leaf mat; wood frogs and toads merge with their surrounding browns; white-throated sparrows, ovenbirds, and ruffed grouse secrete their ground nests. Here grow the finely moulded and delicately tinted of the forest flowers: Bishop's cap, sandwort, dewberry, woodland strawberry, and white violet—now being succeeded by mayflower, bunchberry, the pink-tinged, pendant funneled twinflower, and the nodding, cupped pink and white wintergreens. An aged morel, its cap blackened and melting like plastic, is covered with little sepsid flies who display their light, pirouetting walk and asynchronously waving wings. Under the double protection of tree and shrub canopies this ground level is briefly a world apart, receiving green-filtered sunlight and leaf-dripped rain.

By late morning white down again floats through the air, now from balsam poplars. The aspen seed fall ended a few days ago and the dry-shelled catkins now clutter the trails and catch on shrubs. But this second storm is light. In years when aspens produce in excess, balsam poplars usually yield a light crop, and vice versa.

I begin to feel the heat. Bird song is tapering off. Swallowtails flutter everywhere visiting a plethora of sweet flowers. Frequently they alight on rose, and in juxtaposition the brilliant hues pulsate. I disturb a small forester moth, but it soon lands on a leaf with its wings wide open—eight white blotches, two on each wing, are set in velvet black, and two orange dashes angle forward on the thorax. Another small moth is even more conspicuous, but for a different reason. In flight its steady, rapid wingbeats, although its forward progress is slow, give the impression of a rotating wheel. When it alights on bark however, it is almost invisible, and I see that the dark-light grey bands arc across the wings of each side in opposing semi-circles. Many chloropid flies glide through the foliage. They are under a quarter-inch long, but easily visible because of their cylindrical gold and black striped abdomens. Slender gleaming black ichneumons, their legs and antennae annulated midway with yellow, frequently land on leaves, advance with jerky steps and all the time wave their long antennae up and down tipping the leaf surface in their search for insect hosts on which to lay their eggs.

Near where I flushed the broad-winged hawk several days ago I am assaulted by an alarm: *ip-eeeeeeeeee*, the piercing, trailing cry comes again and again. For many minutes I search the trees, then push slowly towards the source of the cries. Eventually a dark form flaps away from a branch, one I had already passed over without seeing the sun-and-shadow-dappled body of the broad-winged hawk. But it does not go far and soon its screech comes again, now half behind me. The nest must be near, but I can find nothing positive.

I walk on, green lacewings fluttering ahead of me. Their prey is now bountiful. In addition to the first green aphids I find colonies of red ones on the tips of young goldenrod stems. However, few of either group have yet produced the winged individuals which will disperse and pioneer new colonies throughout the rest of the summer. For the present the rapid growth of each colony results from adult females giving birth to live young asexually. In a few days these offspring, also females, mature and do the same. But keeping pace with the increase, and often gaining as they themselves grow, are many ladybird, lacewing, and hover fly larvae. Invisible, but no doubt also present, are internally parasitic larvae which consume all except the skins of their hosts. Prolonged wet cold weather, if it comes, will further suppress the aphids. On the other hand, ants by attempting to repel intruders tend to counterbalance predators and parasites, although their effectiveness is far from perfect.

I chance upon more conglomerations of spiderlings, all very young and enmeshed in silk that pulls shrub leaves close in partial shelter.

I have not gone far from the last of these when I hear a loud whining ahead, close to the ground. Then an eruption of feathers and beating wings blocks the trail. I glimpse a scurry of downy blobs just before the disheveled apparition opens its beak and hisses like an angry cat. Confronted by an aroused ruffed grouse hen I am surprised enough to take a backward step. She advances still fluffed out, hissing, whining, wings whirring. I retreat farther, until she is almost hidden. I must be to her too, since she slowly contracts like a deflating balloon. For another minute she stands still facing me, but silent now and her plumage smooth except for a raised comb of crown feathers. She jerks her head from side to side looking in all directions for danger, then turns and retraces her way with mincing steps, head held high, turning often to look in my direction.

As I wait for the grouse to reassemble her chicks and lead them away from the trail a female hairy woodpecker flies to a dead poplar trunk and begins exploring it for food. She hitches up several inches at a time and stops frequently to tap. Sometimes this becomes sustained drilling and pale wood chips drop away from the tree. Then the bird moves to another dead poplar, swooping up in landing. After progressing upward several feet she reverses, lifting her tail away from the wood so the pointed feathers do not catch, yet working downward as easily as up. Soon she again reverses and spirals toward the top of the tree.

A bird arrives in a nearby sapling, distracting me from the woodpecker. I focus binoculars on a male rose-breasted grosbeak. He must be a yearling since his black is overtoned with brown, and the pink breast triangle is incomplete. Nevertheless his song is true. He moves about from branch to branch inspecting the foliage and singing intermittently. Soon he flies up into the tree crowns continuing his search for insects.

The trail gradually nears the lake; a plain of water glares back at the sun. The forest thins and undergrowth bulges against my legs. Snowberry flower buds are appearing in the leaf axils and the first Philadelphia fleabanes are blooming, yellow centres rayed with pink often so pale it looks white in full sunlight. This fleabane is the earliest flowering of the native composites, for the majority bloom late to signal that summer is waning.

In dead poplars at the edge of a bay, tree swallows fly to their nests carrying food. In the cattails below, both male and female red-wings skulk with insect-bristling beaks. Crows yelp as two harry a red-tailed hawk across a corner of the lake. The hawk is bothered by the big black birds flying close about it and dodges and stumbles as though encumbered by a wind-whipped cloak. The crows pursue and all three birds disappear in the trees, but I can hear their hoarse cries for some time after.

Aware of my coming, ducks swim away from where they were concealed by the cattails—drakes resplendent in burnished nuptial plumage going their own ways; hens trailing lines of ducklings, some of which, a third grown, are losing their baby chubbiness although still downy.

A pair of red-necked grebes occupies a small cove in solitude. One is composed on the nest, black-crested, white-cheeked head sunk down, yellow dagger beak horizontal. Nearby its mate preens for several minutes. Without strain its long neck arches, twists, reaches sideways, over the back and down to the tail, the pointed beak probing every feather. So much of a bird's skeleton is fused and rigid in the interests of flight, that all flexibility seems to be concentrated in the neck, giving it a boneless freedom. Even the legs and feet, used for walking, running, perching, swimming, scratching, or grappling are not much more than highly specialized struts. White flashes unexpectedly as the grebe half rolls to dress its belly; unexpectedly because there is no hint of white above the waterline when the bird is swimming. Adults of a number of loon, grebe, duck, and goose species possess the combination of dark sides and light or white belly, while young waterfowl are universally light below. This probably renders them less visible to underwater predators.

I sit on the bank looking out to the islands. A lone duckling not more than a week old swims along the shore. It dips its beak frequently into the water and at the same time utters an almost continuous peeping. I wait; but no other ducklings appear, no mother. This one must have become separated from the rest of its family and is lost. Although it can feed itself adequately, it is probably too young to survive on its own and will soon be a meal for a mink, crow, or gull unless it joins up with another brood—some hens will adopt orphans, but in general they repel strange young ducklings. For once mindless of leeches and swimmer's itch, I take off my boots and wade hopefully toward the duckling, my plan being to keep it a few weeks until it is old enough to fend for itself. However, in spite of soft tones and slow approach my intent is misunderstood and I only succeed in terrifying the little creature who thrashes away faster than I can walk. It is better to leave it alone and hope for the best. But emotionally I cannot reconcile the fate of the individual with the fate of the species. I know that on this lake alone many ducklings will die from various causes during their first month or two of life—the broods of older juveniles are already smaller than those of newly hatched. It has been so for eons, and will continue so. Testing through time has scaled the number of eggs to meet the expected losses before the new generation reaches reproductive age. In benign seasons more young survive and the population increases; in adverse ones nearly all the young may perish, often many adults as well, and the population declines. Nonetheless, empathy always clouds rationality when I am face to face with the individual in distress, unless it is so seriously sick or injured that quick death is a release.

I am still preoccupied with the lost duckling as I start back and nearly blunder into a family of chickadees. The three short-tailed fledglings maintain a continuous wheezy pleading as they hop from twig to twig in dense shrubbery. As soon as a sleek parent arrives with a dangling caterpillar or small pale moth the pleading verges on hysteria, wings tremble, and beaks gape. The female probably incubated at least six eggs—as many as eight may be laid—and if all hatched she has now lost half her brood. But her care is focused on the living. Usually only when an entire litter or brood dies do the parents show acute distress, and then often the physiological response is to mate again almost immediately. Sentimentality must not endanger species' continuity. However, close attachments can form between adults, especially in mammals, where the death of a sibling or mate causes a prolonged sense of bercavement. Ponderous thoughts for a summer day pulsing with life, but death has always been one of man's obsessions, whereas for other animals the present is by far the most significant moment in life.

I come downhill into the cool shade of venerable white spruce. Somewhere overhead a myrtle warbler sings languidly, absent-mindedly. Boreal chickadees whisper deep within the adjacent bog. Marginally a bog is less acidic than its interior, and I peer through a close mesh of alder, willow, red-osier dogwood, waist-high grasses, and thickly ranked horsetails. Just beyond, in sunny spaces between the black spruce, Labrador tea is whitened with blossom. Below its open leafwork I catch the white gleam of cloud-berry stars and the lacy spikes of three-leaved Solomon's seal. I move a little closer and find the rose-madder petals of dwarf raspberry as well as the pale pink bells of blueberry and bog cran-berry. Bumblebees, hover flies, and a few tiger swallowtails jostle for space at the Labrador tea flowers, almost ignoring the others—in the forest similar interest is displayed in wild raspberry.

Farther into shade the clumped slender stems of stiff clubmoss, whorled with simple spiky leaves, rise several inches above the moss. Many are tipped with the beginnings of strobili. Like horse-tails, these plants are of ancient lineage and in the great Coal Age forests included large tree forms. The attenuated tops of the black spruce are clotted with old brown cones, scales parted, seeds long since shed. The young pale blue-green needles, as on white spruce, give the foliage a deliberately detailed presence. Larch seem spindly and threadbare beside the swaddled spruce.

The seep below the beaver dam has so shrunk that it no longer flows, but the ground remains waterlogged and spongy. Golden saxifrage are clearly defined on the dark mud, neat crenate leaves climbing the short stems and gathering beneath the few, tiny flowers that would be scarcely visible were the adjacent leaves not yellow-tinted like the sepals. These plants will change little before frost shrivels them since the passing of their flowers to seed is apparent only after close inspection.

I leave the comfort of the forest and go to Tawayik Lake early in the afternoon. Unshaded, the sun is fierce and unrelenting, the tentative breeze hot. I linger at the forest edge waiting for cloud, but the small, scattered cumulus offer no relief.

Bird movement wavers branches in an aspen grove and resolves into a family of bluebirds—three fledglings and the two parents. The adults are catching insects, snapping them out of the air and dropping onto them in the grass; then feeding them to the young who quiver and wheeze in ecstasy.

I use the shade of the aspen groves as stepping stones to the lakeshore and on the way acknowledge the cream, papery heads of pussy-toes, and the six-pointed, purple flowers of blue-eyed grass. Most butterflies are hiding from the sun and the ones I see are those I happen to disturb: tailed-blues, pearl and tawny crescents, a soot-dark skipper, a fawn ringlet, and a brown wood nymph with, across each forewing, a broad bar of muted orange inscribed by two dark circles.

I come slowly up the shoreline levee behind a tree for shade and cover. A few feet away a spotted sandpiper flushes with shrill peeps, but lands again a short distance along the shore. Ducks and grebes continue to feed and swim undisturbed. All along the land edge, just back from the water, marsh ragworts stand alone in a ragged line, tall white-hairy stems and leaves headed with floppy hats of deep yellow flowers. Some of the plants, broken early in the season, are stunted to a foot or so, but undamaged ones are level with my waist. On one of the near ragworts a Milbert's tortoiseshell is methodically drinking nectar. At a little distance the wide band of clear yellow and orange extending across each pair of wings, and set against dark brown, is its most visible marking. Satisfied, it flies skipping over the other ragworts. These flowers are attractive to many insects. Bumblebees, wasps, hornets, hover flies, small beetles, and bugs all come and go while I watch. Looking more closely I see the dark specks of thrips sliding between the individual florets of a single blossom.

Pondweeds grow far out from shore, their extent now revealed by half-exposed leaves and knobby flower spikes. Over this area and often dipping down to the water a seething multitude of damselflies has congregated for mating and egg laying. I concentrate on one individual. Like others it pursues an independent course, this way and that, rising a few inches, falling as far, rarely hovering still. But soon my eyes weary with the effort of following one in a hundred thousand or more and I take in the whole gathering. In aggregate their darting bodies and glinting wings are a kinetic abstraction. Damselflies also weave over the meadow, but not as abundantly. Here pairs looped in copulation often pause on shrubs. Darner and skimmer dragonflies stack above their lesser relatives, even well above the trees.

I turn along the shore into the molten west. The heat has driven the bison into shade. I continue on their narrow trail along the levee toward an old willow whose main stem has thickened over the years into a distinct trunk. It is perhaps a quarter-mile away. I plod slowly along thinking how much more comfortable I would be with a hat to shield my head and eyes. Suddenly a killdeer flutters at the edge of an old wallow. Its alarm cries ring sharply and its chestnut tail fans as, with dragging wings, it limps away trying to distract me from its precious eggs or downy young. I look briefly without stepping off the trail, but see neither and walk on.

As I reach the bulrushes in the northwest corner of the lake the black terns who had been out over the water converge on me as if I were a magnet. Their screaming dives focusing on my head and veering off only at the last moment are certain indication of a nesting colony. They pursue me to the willow, but lose interest soon after I sink down into the long grass in the shade, negate my human form and am still.

Beyond this tree willow are others like it, all between the lake and the forest, and in one hulks a huge, roofed magpie nest. It is in good repair and twice in the next hour an adult bird arrives and enters, staying inside for almost a minute.

I relax. To the animal residents I now might be an outgrowth of the willow. A clay-colored sparrow comes to the water through my tree with short flights from branch to twig, to lower twig and finally to the ground. It hops into the water as far as its short legs will allow, ducks its head vigorously and scatters drops of pearl over its back with flicking wings. Its shower finished the sparrow hops out of the water again, shakes and moves up into the safety of the willow to preen spike-tipped wet feathers. Not far along the shore a drake blue-winged teal stands on a piece of driftwood preening. This activity is important in the life of birds, far more so than grooming in most mammals, except perhaps for aquatic species as the beaver, muskrat and otter. Preening restores the unity of the feather vanes, which is essential for efficient flight and insulation; it removes dirt, seeds, and some external parasites; and involves oiling the feathers, beak, and legs to keep them waterproof and prevent brittleness (a minority of birds such as herons and doves produce a keratinous powder instead of oil, but it serves the same purpose). The bathing that usually precedes intensive preening, and it can be in dust rather than water, floats off surface dirt. In addition, birds often rearrange disordered feathers by first raising them, then letting them subside slowly while giving a vigorous shake.

This year water has risen shoreward from the bulrushes and sedges, and is now crowded with the aerial, yellow snapdragon flowers of bladderwort. The rest of the plant, floating near the surface, is a pale concoction of wiry stems and filamentous leaves which bear numerous lidded bladders. These give buoyancy, but also are traps—when nudged by a mite, water flea, Cyclops, or some insect larva of equal size, the bladders open and the animal is sucked in. There its breakdown supplies a much needed source of nitrogen and other nutrients, although the plant does possess a little chlorophyll as well. In autumn the bladderwort disintegrates leaving only a fat, leafy bud about an inch long that sinks to the bottom.

As I sit drowsily, red-wings and the odd cowbird drop down to the hoof-pocked margin to drink; grackles come to drink and also gather insects for their nestlings. The bulrushes cloak a continuity of bird sounds—red-wings' whistles and occasional songs, coots' petulent croaks, ducks' muffled quacks, and intermittently a sora rail's tremulous whinny. Beyond, the uneven conversations of hundreds of ducks and grebes are punctuated by exclamations from black terns.

The sun is well past its zenith, but still six hours to its setting, and feels hotter now than at mid-day. I leave the willow. Walking, I immediately arouse the terns' ire and am again subjected to their squalling dives until the colony is well behind. The meadow seems endless with the sun baking the back of my head and my useless shadow flat before me.

The solstice itself is anti-climactic, subdued, and daylight abbreviated by a low, amorphous overcast from which drizzle softens down all day glazing everything uniformly, yet without the dripping accompaniment of rain. Yesterday afternoon there was brief fury. Early in the day cirrus trailed across the sky ahead of the front. These soon sheeted into alto-stratus, only to be obliterated by a succession of squalls driving down half-inch hail and curtained rain for most of the afternoon before relenting and sweeping on, leaving a legacy of cloud and persistent wetness. I feel cold—in less than a day it has dropped forty degrees out of the low nineties—and wear a sweater under my poncho. But bird song has revived and chorus frogs sing again, although raggedly and without their springtime fervour.

The far shore and islands are faded, their colours greyed. Waterbirds flock the smooth water as dark blobs—singly, paired, small groups, and broods. A few gulls and terns shuttle back and forth over the lake, structuring the empty air.

I turn away inland through forest and past several ponds. In this light the constituents of the undergrowth regain their identity, leaf shapes no longer disrupted by blotches and spatters of sun and shade which meld the dissimilar. Flowers too reveal subtlety of form and colour now they are not divided by hard shadow lines or bleached by strong light.

The rain has laid the dust and dissolved many fragrances that even my dull human nostrils can savour, allowing me to skirt the periphery of a sense realm that to most of our mammal relatives is as varied and vivid, and replete with nuances, as colour is to us. I know when I pass a balsam poplar or a spruce—poplar resin is tangy, that of the spruce slightly acrid. I can identify rose as a delicate perfume when I bend toward several flowers. My nose wrinkles at the strong, pungent-sour emanation from cow parsnip heads. But my limits are soon reached and for the most part unless

I touch a blossom with my nose I can only separate wet earth from wet leaves from wet bark. I am almost as "scent blind" as some mammals are "colour blind."

A young robin flutters up from the trail and alights on a low branch. It is still heavily spotted below, but almost adult in size, posture and general plumage; when it flies its movements are sure.

Moments later a sapsucker dips from an aspen, but instead of silence in its wake a dry, hoarse trilling radiates through the tree. On the southwest side of the trunk I find the round entrance to a nest cavity about twelve feet above the ground. Perhaps because there is little to fear from predators in such a retreat—the hole is too small for a red squirrel to enter—the nestlings call continuously even when the parents are absent. In a few minutes the male arrives, abruptly clapping onto the trunk just below the entrance, and pokes his head and neck in up to the shoulders. On his arrival the young voices rise to a crescendo of excitement. The adult soon leaves again however, off on another food collecting trip.

Near a pond a house wren sings with boundless enthusiasm, and I locate him in a small willow. As I watch, a second wren who must be his mate enters the shrub and pleads fledgling-like with quivering wings and gaping beak. The male appears to feed her, then moves off and resumes singing. She hops away on her own and is soon out of sight.

Dabbling ducks feed in a deep bowl recently dammed by beavers and now ringed with the grey ghosts of drowned trees, and paved with duckweed. A pair of widgeons, swimming close together, meander across the centre. They dip their beak tips through the duckweed and slide them briefly along the water surface before raising and swallowing. They leave behind a dark, snaky trail that green soon floats over, but does not entirely obliterate. Nearer, a drake gadwall paddles between standing and fallen trees, gradually coming closer and feeding the whole time. He does not tip-up, but like the widgeons skims his beak horizontally and occasionally plunges it down to the base so that when he lifts it again a garland of duckweed clings to his feathers. At intervals of ten to fifteen seconds he utters a single, hard *kek*, for which there seems little purpose since the bird does not appear to expect a response. The widgeons are silent as are two hen mallards and a pair of green-winged teal at the far side of the pond. Casual observations can be tantalizing and perplexing, raising many "why's" that sometimes are not answered by later consultation. If I could come often enough or could stay long enough at this place I might find the reason for this gadwall's calling. No one has yet answered my "why."

Now at the end of June I sense vaguely, but would find difficult to measure, a subtle change in the summer's tempo—less emphasis on the acceleration of becoming, more on the maturation of what has become. Half-grown fruits and seeds have replaced early flowers; some nestlings have become fledglings, and some fledglings independent juveniles; ducklings are growing from down into feathers. And today I also find premonitions of summer's end in the first young aster flower buds, in the points of next year's leaf buds in the axils of existing leaves. Past, present, and future merge in the tender new leaves appearing on still lengthening shoots and twigs. In nature time and processes are a continuum. Phase overlaps phase, even death is but a redirecting of elements. Only during low-temperature dormancy is the flow temporarily frozen still, yet this intermission is often essential before further development can proceed.

For some insects plant leaves are shelter, rather than exclusively food. Neat semicircles missing from rose leaflets have been sliced out by leaf-cutter bees who use the fragments to fashion individual cells, each supplied with pollen and nectar, for their larvae. The cells are built successively along a narrow tunnel excavated by the bee in a decay-softened log. Once the tunnel is filled the female leaves her offspring to develop on their own, for these bees are solitary.

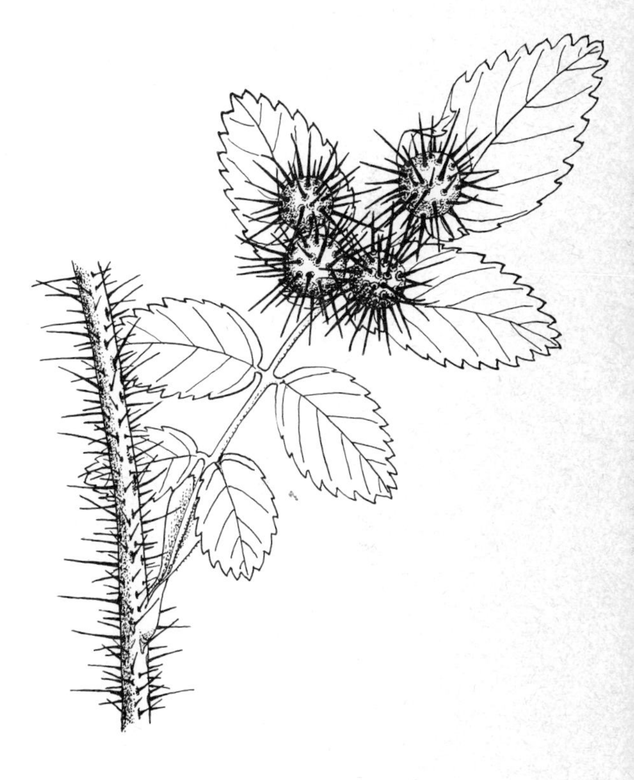

Spiny rose galls are unusually abundant this summer, occurring in small clusters on the leaves. Each gall is a sphere up to a half-inch in diameter and resembles a red sea urchin. The origin is a miniscule wasp larva whose secretions redirect cell growth to produce the abnormality, yet one which has a constant form. The rose galls are only one of hundreds of different types caused variously by mites, aphids, sawflies, a few moths, and true flies in addition to the tiny cynipid wasps. The majority of gall makers are inconspicuous as adults and pass unnoticed except by the people who study them; the misshapen plant parts in which the larvae feed and mature are the obvious aspects of these insects' activities. From where I stand I can see many red-pimpled aspen leaves; a willow twig bearing the mimic of a multiple rose flower in its altered leaves; white spruce twigs tipped with pale cones, each false scale needle-pointed; and pregnant swellings on goldenrod stems.

Leaf miners, whose larvae complete their development in the stable, relatively secure environment between the upper and lower epidermis of a leaf, are another diverse and abundant group, dominated by flies, but including small moths, beetles, and a few sawflies. The eaten part of the leaf, usually either an irregular linear channel or a patchy blotch, shows clearly by its pallor or discolouration.

By mid-morning the heat has seeped deeply into the forest, although here and there I sample the delicious humid coolness of lingering night reservoirs in hollows and in the shadow of spruce. Bird song evaporates with the dew and as the day progresses becomes more and more intermittent, with the exception of a few irrepressibles—house wrens, song sparrows, Tennessee and yellow warblers, red-eyed vireos, and least flycatchers.

An oriole perches overhead. He is dull for a male. Most of the feathers on his near black head are broadly edged with subdued orange, giving a scaled effect; his underparts are the same tone of orange, but unmarked, and pale to yellow near the tail. The bird is too high up to see any more of it, but he must be a year-old adult male who has failed to moult into nuptial plumage, as it is too early for this year's young to be full-grown and independent. The oriole chatters briefly and flies off. He is alone.

A flicker arrives and perches crossways on a branch of a dead balsam poplar, a habit more common in this than most other woodpecker species. In a few moments the bird, a female, drops to the main trunk. Immediately hoarse wheezing erupts and two well-feathered flicker heads poke out of a dark hole about fifteen feet above the ground on the northeast side of the tree. The adult feeds her young by regurgitation with jerky, violent movements that seem satisfactory to both, then squeezes into the cavity probably to feed other nestlings. When she emerges she flies fast and direct through the trees. The nestlings fall silent and do not reappear at the hole.

White peavine has now been almost completely replaced by the very similar purple peavine. The young flower is a uniform magenta-purple, but with age the keel pales contrastingly. Meadowrues, strawberries, twinberries, and low-bush cranberries are following the white peavine into oblivion. Although still flecked with buds, rose thickets are now embellished with many five-pointed stars, the pale green sepals revealed when the petals fall. As the fruit begins to enlarge they will bend forward and converge. Red-osier dogwoods, high-bush cranberries, yellow avens, raspberries, cow parsnips, Canada anemones, and snowberries bloom profusely. Under them and among the bunchberries, twinflowers, mayflowers, dewberries, and wintergreens, I find the occasional retiring Habenaria orchid, green of stem, leaf, sepal, and petal. Most bedstraws are weak, sprawling plants with a few, very small greenish-white flowers, but the northern bedstraw is sturdy, erect, from one to two feet high, and its bright white flowers are concentrated in a foamy terminal cone. Many are lighting today wherever an edge exposes them to sun. In the same locations fireweeds and dogbanes are thickly budded, while the first orange wood lilies fire the open slopes with flamenco brilliance.

I come down to the edge of a bog, cross the squelching moat, and stop. In wet bogs boots soon break and crush the delicate living moss and churn the brown peat into a thick sludge. In time, if left alone, the scar heals, but mosses other than Sphagnum are often the colonizers and the original character of the community is altered. I look ahead. Labrador tea, blueberry, and cloudberry blooms are fading, yet many three-leaved Solomon's seal, bog cranberries, and dwarf cranberries remain. The last are distinctive with their long narrow recurved pink petals and exposed anthers fused in a tube around the slender stigma. The trailing, small-leaved stems are overgrown with moss.

A family of boreal chickadees comes close from the depths of the bog. The two young are almost grown, but their feathers are still fluffy and their markings blurred. They keep close to their parents and frequently utter husky *day-day*'s. Farther in a red-breasted nuthatch pipes thinly. With short, hollow-toned calls a gray jay flies to the top of a spruce. Its smoky colour, slightly paler underneath, is accented by a white face, charcoal nape, and black beak, eyes, and legs. This jay nests very early; it is already incubating by the end of March. However, as neither incubation nor nestling periods are prolonged, fledged young are travelling with their parents before most other small birds have even hatched. Yet only one brood is raised each year. Gray jays probably nest in the park, but are rare and I have never seen the dark sooty juveniles.

Damselflies and dragonflies weave bright nets of shifting colour as they hunt. They are far from the pond or lake of their origin and the water where they will lay their eggs, but these strong insects often range widely seeking swarms of midges and mosquitoes. A few butterflies flicker among the spruce, most having been attracted elsewhere by more appealing flowers than the bogs now offer. A grey robber fly comes to rest on a leaf beside me and stands there for some time twisting its head with surprising mobility. About half an inch long, it is all eyes in front, long slender abdomen behind. The heavy legs bristle with stout hairs and end in curved claws. In reality this fly is not a robber, but a predator of the first order, pouncing on its insect prey with the vigour of a wolf, piercing it with the curved blade of its proboscis, and finally sucking dry its body fluids. A brush and beard of bristles on the face protect the eyes from injury, for this fly has no poison to calm its struggling captives.

I return to the shady forest and at the edge of the moat encounter five juncoes in the alders. Three are brown-streaked juveniles, out of the nest a few days but still very dependent, although they beg only when a parent arrives with food. The male sings once, out of sight on a high perch.

On willow twigs I count many fat ladybird larvae, some black with white markings, others black with red. Their feeding is more like grazing than capture since the wingless aphids neither attempt to escapc nor struggle when grasped. Other aphid clusters are going down before the advance of lacewing and hover fly larvae. An estimate of predation on fifty percent of the colonies would not be too high. Here and there a few winged aphids are ready to leave and found new colonies.

I make my way toward the kestrel nest tree. Along an old bulldozed track cut into the south side of a hill I pass a coyote den. Nearly two months ago the telltale pile of loose sandy soil at the entrance was imprinted with the adults' paw marks. Now cub tracks have been added, but when I put my head down to the hole, draped over with long grasses, I can hear nothing from within. There is no point in waiting. If the coyotes are anywhere near they will long have been aware of me and will stay in hiding until I leave.

When I arrive at the open slope the female kestrel is perched at the top of the nest tree. Suddenly her mate appears and flies to her. As he nears she starts a run of shrill *ki*'s, crouches slightly and vibrates her wings, a supplication that brings him to perch beside her and extend his beak almost to touch hers. But he has nothing in it, and the feeding gesture is no more than that. Soon he flies again, now along the edge of the forest; then I lose him. The female remains and preens carefully.

Standing just inside the forest, I stay over half an hour to keep an eye on the kestrels. The sunstruck slope is a highway for butterflies. Most of the swallowtails appear worn, hindwings ragged and notched, one or both tails missing and coloured scales rubbed off; yet they still fly strongly. However, alfalfas, fritillaries, and pearl crescents are now the common species. The little "blues" have declined greatly during the past week. Once a red admiral flashes past, red-orange sash bands on dark wings making recognition easy. Omnipresent hover flies hang in the air, rest on leaves and feed at flowers. There are several species, the most distinctive being a large hornet mimic, with a hornet's elongated black and white body and long narrow wings. The fly adds to the deception by waving its abdomen up and down as it walks.

An insect lands on my hand. I am just about to brush it off when I see that it is a very strange-looking fly—a tiny black head points down below the convex, lustrous black thorax; the black and white abdomen is large and heavy; but the wings seem barely sufficient for flight. The larvae of flies in this family, the Acroceridae, are all internal parasites of spiders, mostly web-weavers, and the small wings may help keep the female from being snared. However, she does not endanger her own life by attempting to lay eggs directly on a spider. Instead she deposits them, several hundred or more, on a nearby twig or leaf. The mobile larvae find the web and progress, unhindered by the glue drops, along a strand like looper caterpillars. They can also jump a short distance and in this way attach to the spider. Those that survive the spider's attempts to scrape them off burrow through its exoskeleton and imbibe the body fluids. The Gordian Knot life cycles of many invertebrates are intriguing, if macabre in the case of parasites.

I glance at the kestrel tree. It is empty and the slope consumed by the sun. Leaves twisting in the breeze flash reflections as if they were glass. In the glare and heat the forest-walled opening has become a bubbling vat, the buzzing clay-colored sparrows escaping gases from the seethe. Looking in I feel the radiating heat, but beside me, tempered by the shade, the impinging sunlight laps gently on leaves, not too hot for basking insects.

Cedar waxwings trill, pour into the top of the dead poplar and separate into the constituent five birds. Although waxwings have been back for over a month they may not yet have started to nest. The nestlings are fed some berries in addition to insects when only a few days old and nesting is timed to coincide with the ripening of a variety of juicy fruits. These perennially sociable birds often nest in close proximity, then travel to a common feeding ground, or temporarily unoccupied males may gather together, so small flocks can be seen at any time. The waxwings linger on, conversing in sibilants, looking about, preening.

They leave as one with the return of the female kestrel although she does not threaten them. She lands on a branch close to the nest hole carrying what looks like a vole. Two intermediate perches take her to the entrance and then suddenly she is inside. I cannot hear any calling, but presumably she is dividing her kill among the nestlings. Five minutes later she emerges, flies to a nearby branch and, hunching over a scrap held in one foot, soon finishes the remains of the vole. She strops her beak, probes her breast feathers and the underside of one wing, and finally settles into wide-eyed rest.

I quickly cross a small upland meadow. Grass has been cropped, bent, and flattened by grazing and resting bison. The mounds of dung are frequented by hairy, yellow, long-legged male dung flies awaiting the arrival of the smaller, dull grey females. Their larvae are one of many insects that hasten the breakdown of mammal excreta, but the adults are predatory. The bison have eaten only grasses, leaving the flowers untouched: clusters of curd-headed pussy-toes whose short simple stems arise from a mat of small tomentose leaves so dense that they inhibit the grass, scattered yarrow each crowned with a mass of white flowers just opening, dandelions that are yellow periods or airy white seed balls, a few blue-eyed grass and occasional leafy stands of grey-green sage-worts, prickly thistles, and smooth meadow fleabanes.

At every pond and lake I visit today I find drakes beginning their midsummer moult. Scattered dull feathers already mar the nuptial hues. Gradually over the next three to four weeks these birds will come to resemble the hens of their species. By late September a second moult, of the body plumage only, will have returned the familiar nuptial patterns. Grebes and loons also moult twice a year, but the drab winter greys and browns are hardly noticeable before September.

Along the shores aquatic invertebrates are so dense that they form a living soup. On the surface whirlygig beetles spin in dizzy, black-moleculed groups, while water striders, represented now by many small nymphs as well as adults, skate away or stand in spread-legged, water dimpling posture. Snails glide on every submerged plant and occasionally on the underside of the water's surface film. Some have long tapering shells, others, in a different family, are whorled. Many are united in pairs for the mutual transfer of sperm. Water boatmen and backswimmer nymphs mingle with a lesser number of adults. I see larval as well as adult water beetles, wriggling four-legged tadpoles, and dark undulating leeches. A piece of twig on the bottom twitches as the head and a

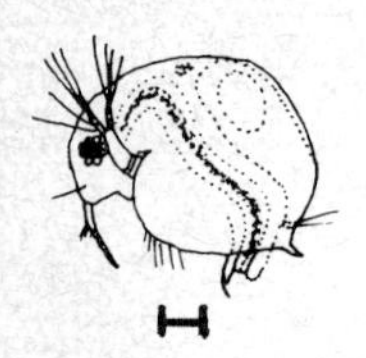

water flea
Cladoceran

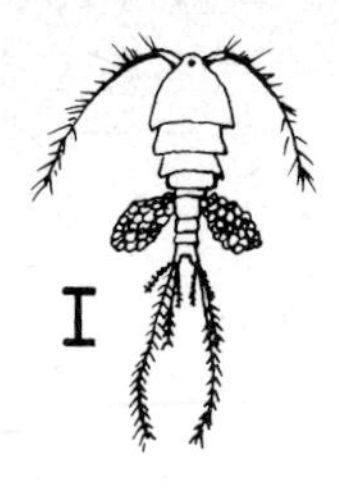

Cyclops sp.
(♀ with eggs)

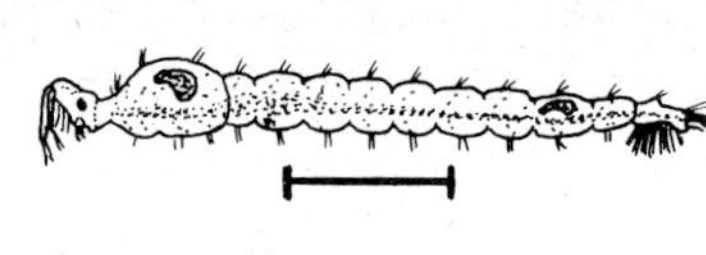

phantom midge
Chaoborus sp.

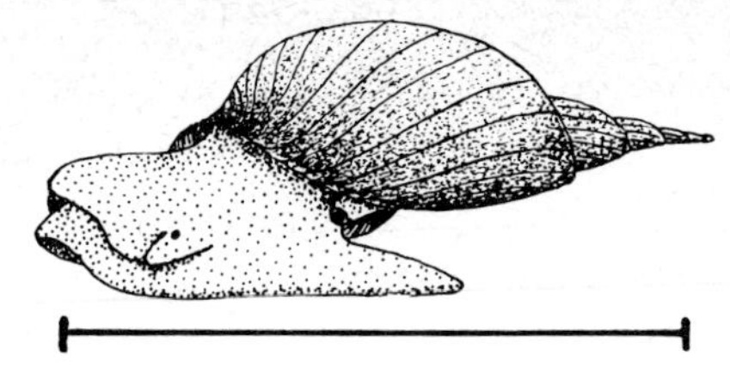

pond snail
Lymnaea sp.

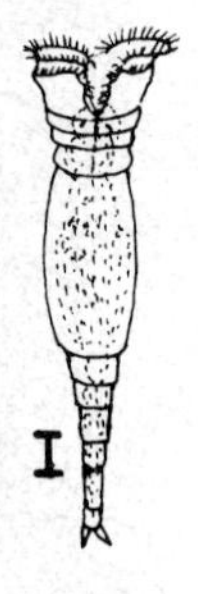

rotifer

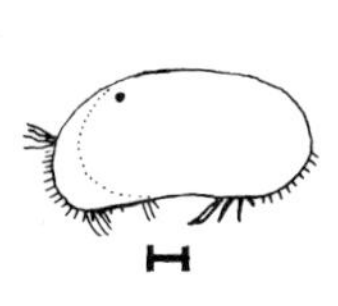

ostracod

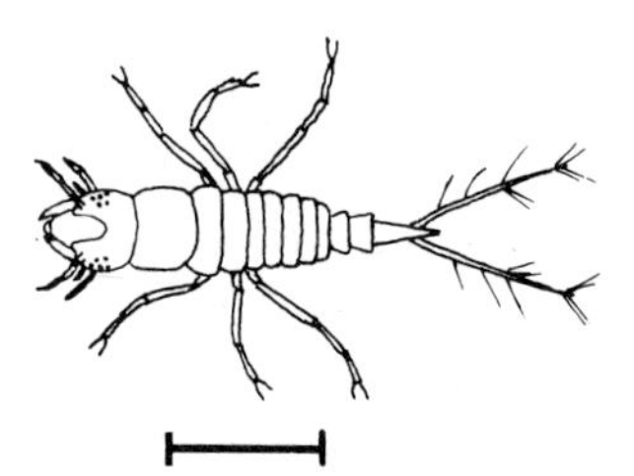

diving beetle larva
Hydroporus sp.

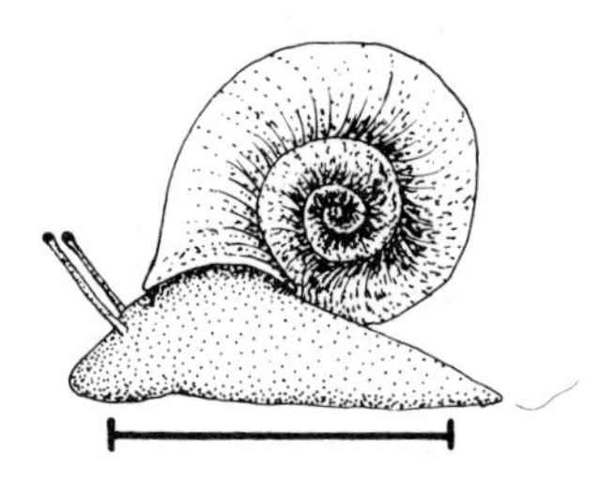

orb snail
Helisoma trivolvis

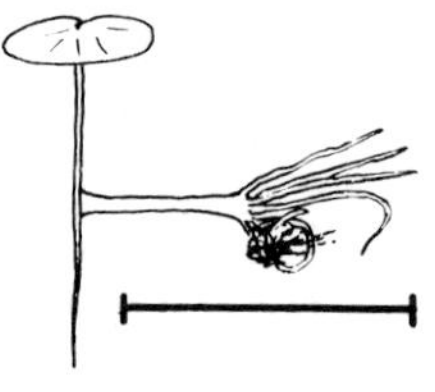
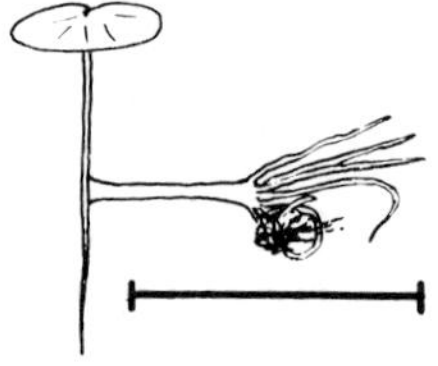

Hydra sp.
(with prey)

Paramecium sp.
(microscopic)

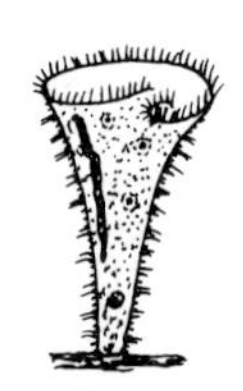

Stentor sp.
(microscopic)

pair of legs belonging to a caddisfly larva emerge and it slowly crawls along. Shrimps scull away, many in mating duos. Perhaps if I watched them for longer periods I would discover greater variety in their behaviour. I dip a jar in the water and bring up myriad lives I did not see from my position above: clear colourless water fleas and cyclops, mud-brown bivalved ostracods, red mites and, anchored to a duckweed root, a milky hydra with its tentacles wrapped around a cyclops. I recognize a feathery-gilled mayfly nymph; midge larvae—including a single specimen of the transparent Chaoborus, or phantom midge, which normally floats in a horizontal position with the aid of dark-pigmented hydrostatic organs fore and aft; the jellied, white polka-dotted glob of a snail's egg mass, and a long, slender writhing aquatic worm. Dots and hairs of green are colonial and filamentous algae. Under a microscope the algae would become intricately designed cells, and additional life forms would appear: rotifers, flagellated protozoa, ciliated protozoa such as Paramecium, Stentor, and Vorticella, and copepod larvae.

Lake waters are green with the first major algal bloom of the summer, temperature, nutrients, and light all having combined favourably. A week ago there might have been fewer than a hundred cells in a teaspoon of water, now they number ten thousand or more. Almost as suddenly, with a change to cool cloudy weather or the exhaustion of essential nutrient, a mass die-off can occur. The water clears, but the ensuing decay of the algae may deplete dissolved oxygen and suffocate many fish. Their decaying bodies in turn foul the water and in shallow lakes it may take weeks for the former balance to be restored, more than a year for the fish to recover.

As I pass an open bog with spruce standing about like sentinels a young snowshoe hare darts away. Except for an initial glimpse I never see it clearly as it bounds off through cover, a dark body low to the ground. The hare, or my movement, or both startle a junco into ticking alarm notes. But as I remain still and the hare stops somewhere and crouches again, the bird's nervousness subsides and it falls silent. Then twigs shake as it shifts around in a willow. A minute later it flies to a spruce and I see that it is a male. As he enters the spruce familiar fledgling wheezing begins. then the junco stuffs a caterpillar into the cavernous gape of a nearly full-grown cowbird almost twice his size and bulk. As soon as the food has been accepted the junco flies off to find more and the cowbird sinks complacently on its perch. This is one of only a few observations I have made here of foster parents to cowbirds, although small sparrows, vireos, and warblers are the most frequently chosen hosts. One ornithologist has suggested that agricultural clearing has greatly encouraged the spread of the cowbird, previously restricted to natural prairies and parklands, and has thus exposed many forest birds to its parasitism. Because the contact is recent, behavioural adaptations have not yet evolved to cope with the problem. Some species which have been closely associated with the cowbird for millenia have adopted specific tactics; for instance, the yellow warbler, yellowthroat warbler, red-eyed vireo, warbling vireo, and alder flycatcher often floor their nests over and start a new clutch of eggs if they return to find a cowbird egg added to their own; and the robin, oriole, and kingbird may build a new nest or simply oust the intruder's egg. Once hatched the fast-growing cowbird nestling dominates the brood so that, sometimes, none of the rightful young survive, although more often the cowbird replaces only one or two.

White-spotted forester moths have recently become common and hurry away from shrubs as I walk along the trail. Here too I encounter a newly emerged white admiral, immaculate, precision made, the five orange spots and pale blue crescents between the white band and the wing margin showing clearly. Fresh adults from now on come from eggs laid early in the spring by overwintering females, and will form this year's hibernating cohort.

On a buffaloberry leaf two coleophorid moth larvae graze like miniature caddisflies. Their dark brown cases are smooth, tubular, and slightly curving toward the tip. The open mouth of each case is attached to the leaf surface where the larva has cut out a hole the same size. To feed it reaches forward, munching on the soft cells between the upper and lower epidermis until it would have to leave its case to go farther. Then it moves to a new site and repeats the process. When the larval phase is complete it will crawl to a place where the case matches its background and pupate inside. A few beetle larvae are the only other known land insects to produce external cases, but they do not mine leaves. The case and mining combination is limited to two families of very small moths.

The sun sinks behind the forest rim. Midges and other small flies collect in dancing swarms, barely visible ghosts in shadow, golden motes in the last shafts of sunlight. Ducks paddle out from the cattails to feed, leaving meandering, interlacing trails through the duckweed. An old beaver lodge stands away from shore, almost like a natural island with its topknot of nettles. Each year as the lodge is freshly packed with mud from the bottom it becomes a richer seed bed. Bird song ignites again, soon flaring into the forest and through the red-wings in the cattails. Tree swallows swoop among erratic dragonflies without collision. One that I follow to its nest is greeted by an almost fledged nestling stretching out for its share from the parent's insect-crammed beak.

A shadow tide rises up the eastern trees, soon flooding and spilling over their crowns. Close to the ground the air is already cool and damp.

July

I sit by the edge of the lake astride a fallen poplar angling into the water. The western sky stays bright and pale long after the sun has gone. The east darkens degree by degree, slowly becoming a violet-tinted grey that is neither opaque nor translucent. My eyes seem to penetrate beyond the limits of the universe, yet there is nothing to see; it is too light for stars. The only concrete object anywhere in that empyrean void is the pasty disc of the moon.

The day's stiff breeze is dying; waves are subsiding to ripples that smooth when they meet the outer edge of the pondweed forest.

I close my eyes briefly and the plastic of visual substance is replaced by transitory strands of sound. These I attempt to use in my restructuring of physical space. The settling-in calls of gulls and terns come from in front, somewhere in the middle distance. Overlapping them however, are the songs and notes of many other species, a motley, disoriented array when I listen to them in concert. To position them in relation to myself I find I must focus on each species separately. I select the red-necked grebes first because they are so numerous that it is easy to relate diminished volume to distance. In this way I line out the lakeshore stretching south, the bay on the west side of the point behind, and the lake's north-west prong. The north and east limits are more difficult as grebes there are fewer and so distant that they are faint and impossible to place precisely, and as well are too often overcome by nearer calls from other grebes. My ears suggest that the lake may extend for miles in those directions. A bittern pumps, off to the northwest. He may be beyond the grebes, but I cannot be certain since his voice is so different from theirs. Various ducks mutter conversationally in an arc sweeping almost a hundred and eighty degrees in front of me, well out where the water is deep.

I start involuntarily at a penetrating *weep-weep-weep* that is close enough to touch. The spotted sandpiper may have landed on my tree at the edge of the water. I strive to pick up sounds indicating the bird's movements, but hear nothing. Possibly it just called in passing flight. A fast flying duck, probably a goldeneye, comes up on my right parallel to the shore. It seems to approach very swiftly, then fade away more gradually.

A snipe winnows high up; red-wings sing and whistle from the cattails fringing the bay behind; a pied-billed grebe and a sora rail call from the same area, yet sound much closer. Forest birds sing on the point and I believe I can tell those in the tree crowns from those in the undergrowth. However, I can never completely isolate all these sounds from my visual knowledge of where to expect them; the connection is too ingrained. The *terra incognita* of unencumbered sound can only be fully explored by the blind.

When I reopen my eyes everything is much dimmer. The western pallor has contracted; the east is more strongly violet. Landscape colours have darkened except for those already pale. They absorb the last sky light and gloom starkly—white birch and aspen trunks, and white flowers beside me, more prominent now than in daylight.

The moon begins to glow. Night has claimed the forest and shapeless black hovers behind the outer trees. Yet here in the open, with the west at my back, I can still distinguish the furrowed bark of balsam poplars, the identity of near ducks, the shapes of leaves. Two bats flutter against the blank sky.

Suddenly I become conscious of the cooler air, humid and redolent with the spice of balsam poplar resin. Exuded in brown stains on the underside of the leaves, it is more potent than most flower scents.

One by one the songbird flames gutter and die out. Last to go are robins, white-throated sparrows, song sparrows, and a Swainson's thrush. Cedar waxwings fly over trilling, but I do not see them. One final red-wing silences and leaves the marsh to the grebes, rails and snipes, a few frogs and toads.

Stars shine in the east and slowly the constellations emerge and brighten. As always the handle of Ursa Minor beacons north, and not far away is Cassiopeia's precise "W." Cygnus, more of a dagger than a swan, lies along the Milky Way. I find a few others, but my favorites, Orion and his constant companion have temporarily sunk below the horizon on some long, celestial hunt.

By midnight the moon is brilliant, a rival to the sun. But its light has strength only at its origin and where it strikes an object. The sky remains dark, depthless and deep, shading lighter only near the horizon where trees etch inky cut-outs. The lake beyond the molten globe reflection of the moon and bright points that are the largest star is as stygian as the sky. However, very close beside me I can see more detail now, even colour, than I could in the earlier dusk. Yet, like the dark-light sides of the moon, cast shadows are absolute.

The air is utterly still; the lake without a ripple. But behind in the undergrowth suggestive rustles breathe and murmur. Nocturnal beetles and spiders scuttle; caterpillars chew on leaves; deermice and shrews move about in search of food, perhaps sometimes in play. Individually few of these sounds would reach my ears; together they are cumulative and are just audible. Occasionally I hear a heavier, purposeful tread suggesting a weasel, or possibly a snowshoe hare.

I am stiff and chilled after three hours of sitting, but seconds before I change position a black shape slides out of the water and onto the poplar trunk. At first it seems to be a coil of the water itself, eel-like and vaguely repulsive. Then I realize it must be a mink. The animal pauses briefly, then pours back into the water.

I eat part of my snack and drink half a thermos of steaming cocoa; stretch my arms and legs. But it is not enough. I climb off the tree and make my way to the track that leads to the end of the point. My passage through the undergrowth obliterates all other sound and I feel like a bulldozer bringing crashing destruction; but once on the track I can move silently. The short walk warms me and I sit again when I can go no farther. Out towards the middle of the lake ducks continue to mutter, occasionally interrupted by a gull's petulance or a red-necked grebe's stentorian bray. A horned owl hoots, so far away that the tone is as soft as a dove's.

Perhaps awakened by my disturbance, or the light of the moon, a white-throated sparrow sings, once. On their own, cowled by the night, the fluted notes attain transcendency. I yearn for the bird to go on and on, yet am half relieved when it does not. Subsequent songs might not be perfect, or my ear might weary. A few minutes later a song sparrow delivers his more complicated, yet less musical refrain. Nevertheless, soloed in the entrancement of the present it achieves a rare quality. This song is not repeated either. A robin and a Swainson's thrush call softly several times;

then all forest birds settle back to deep sleep.

Now I pick up the threads of squeaking bats whenever their erratic flight brings them invisibly close.

The beavers come out—splashings near the lodge, groans and whines, teeth slicing into soft poplar, tearing, ratchety; swish, crash the sapling is down; leaves torn off and chewed with heavy munching. The moon is behind me now so I do not have its distorted reflection to tell me where a beaver is swimming. Suddenly, close enough to make me jump, the alarm slap of a tail cracks on the water. All beaver noises cease instantly. But soon they resume tentatively; then as before.

One hour past midnight my watch becomes a gruelling vigil. I am tormented by the desire for sleep and cannot think of anything else. My body is cold and cramped again. I put on a sweater and jacket; eat and drink the remainder of my food; flex my limbs. I rally briefly, then am sleepier than before because I am more comfortable. I succumb and curl up in my blanket against the log. My head spins and lights pop behind my closed eyes; the earth reels, retreats; the braying grebes fade as I speed into the underworld . . . .

A low moan swells and flows through me—a lost spirit in the Shades? my lost spirit? the last wolf? It comes again—low, protracted, drawn to a fine taper before ceasing. I open my eyes and am awake. The call is real, floating in over the water from a solitary loon. But it is easy to imagine a wailing ghost, one of the loon's ancestors or mine; easy to slip back empathetically into the millenia before Christ, to the era of the Lascaux cave paintings, and beyond. Have we really changed so much? still plagued by the eternal questions—who are we? why? immortal or not? where are we going with all our knowledge, or lack of it? where am I going? Perhaps we are more troubled by these wonderings now when so many of us have cut ourselves adrift from the certainty of faiths or mythologies. Enough, I cry! Back to reality, ego over id, mortal loons of feathers, bones, and blood who will die, decay, and recycle in molecules.

The brief sleep has revived me and I sit up blanketed, alert. After the loon ceases its moaning a general hush like a palpable presence hovers overhead pressing down lightly. Nothing at all stirs on the lake. The beavers have gone into their lodge. The forest is still, the air cold and damp. The moon has sunk low and its weakening radiance deepens the night.

Wavelets of cloud drift in from the west. Soon they dim the moon and everything around me is black and sightless.

The paling east gives the first hint of the returning sun. Then the clouds become visible grey. The stars dim away. Overhead the zenith is dusk again. Water separates from the denser, darker land, and the serrate trees are unquestionably of the earth. They grow trunks, and limbs, and foliage. Their leaves shiver as the air stirs.

Bats returning to roosts flicker overhead. The white-throated sparrow sings and repeats sublimity; sings again a minute later. Red-necked grebes rouse and their braying echoes on from pair to pair. Gulls keen. Ducks converse and a hen mallard in the bay behind engages in a long quacking outburst.

The clouds whiten, reflecting the now bright east. A robin calls softly many times, then sings. Song sparrows, least flycatchers, a chipping sparrow, and the Swainson's thrush join in. A snipe begins winnowing. Red-wings are a scattering of whistles and rough, dry *check*'s.

The sky is more blue than grey, and tender pink tints the clouds. A catbird wakes himself with discordant squeaks and clucks; an oriole chatters. Gulls begin to leave the lake, flying low and steadily northwest. A few ducks skim the water, then land in furrows of white foam. On the shore, only a few feet away, a spotted sandpiper suddenly materializes and feeds. It must have stepped off a nest; I would have noticed had it flown in. The bird is silent and preoccupied, jabbing its beak repeatedly into the water. Crows caw over on the north shore.

The morning blooms. It is full daylight. The clouds fire up vividly. Orange washes up from where the sun will momentarily appear. The red-wings, catbird, robins, and oriole flower into full voice. Yellow and Tennessee warblers, a red-eyed vireo, a house wren, and a rose-breasted grosbeak add substance to the quickening lattice of song. A jay screams twice, tearing gaping holes in the harmony, but apparently only to my ears as the singing does not falter.

The breeze fans the sun and it rises seething, too glaring to meet eye to eye even before it clears the far trees. Its touch is instantly warm on my skin, welcome now. But I can no longer hold back the clamour of my mind and body for sleep. I go down the track stumbling on twigs and pebbles, to a rest of oblivion.

Under a flat ceiling of stratus overcast the early morning reclines heavily, the night clinging stickily, a shedding skin that refuses to slough off. The lake is molten smooth. Reed Island is a distant green fringe, almost a mirage. Leaves are poised, clearly excised from the murk behind. All surfaces are dry, for the warm night was without dew. Yet many insects are sluggish, waiting for sun heat. Birds sing raggedly. A snipe thrums like the sound of a coming wind; but the air is still. Ducks and grebes, gulls and terns swim and fly double-imaged, their reflections scarcely less real than the birds above.

For several minutes I watch a pair of red-necked grebes with two downy young about a week old. The dark and light striping on head and neck makes the chicks conspicuous, although this

morning anything on the water stands out like a model on display in an empty room. They dive often, but always surface close to one or other parent. A California gull beats toward the group then swoops low down on them, not once, but repeatedly. If it is trying to separate and attack one of the chicks it does not succeed, for they press close to the adults who point rapier beaks at the gull. Unable to break the grebes' defence the aggressor soon wheels away.

On the far side of the bay two mallard families swim slowly along the edge of the cattails surface feeding. One hen is followed by seven half-grown ducklings, the other by five larger, well-feathered young.

I look across the reach of the lake. Drakes, noticeably mottled, swim everywhere repeatedly billing their feathers and flexing their wings. Many moulted feathers float near the shore, so light and water repellant that they seem to be suspended in the air. I pick up several and try to place their origins, but once isolated most lose their identity. Black, white, iridescent green, and rufous could belong to any of several species. Even this finely vermiculated grey and white feather could be from a pintail, canvasback, red-head, ring-necked duck, or lesser scaup. But here is one in warm cinnamon tones that must have belonged to a widgeon; a chocolate brown feather that could only have come from the head of a pintail; and another in purplish-grey from a blue-winged teal. When the hens moult, their variegated buffs, tans, and browns will be impossible to pin-point. It is not the individual feather, but the pattern of the masses on the bird that confers distinction.

I focus below the floating feathers. Phantom clouds glide through the gloom—schools of stickleback fry which now and then reflect light as they turn in unison.

Wandering along the shore I meet a family of song sparrows in a willow thicket. An adult sounds the warning first before the young join in with thinner voices. Then all four birds flit about in the willows restlessly, flicking their tails and giving me no more than glimpses of a head, part of a wing, a streaked breast. But as I continue to stand silent their nervousness wanes. Finally I see one of the juveniles, adult in all respects except for the somewhat shorter tail and less heavily streaked breast without the central spot. The presence of only one adult suggests that the female may be starting a second clutch. The male will finally leave these offspring when the new set of eggs hatches and his mate needs help with feeding the nestlings.

A male oriole flies through the tree crowns chattering roughly, but does not pause. Nearby a least flycatcher hiccups irregularly. I hear snatches of song that could be from a robin or a rose-breasted grosbeak.

A chipping sparrow flies from a tangle of raspberries and lowbush cranberries, then drops out of sight uttering a continuous stream of alarm notes. I bend over the place where the bird appeared and at once see the neat, round cup of its nest filled with three, grey-downed chicks breathing heavily, eyes tightly shut.

After breakfast I go to the flicker tree. Three days ago I spent over half an hour there. When I arrived the male was delivering food. He left and did not return. During his absence one of the nestlings remained at the hole and frequently thrust his head far out to peer around, revealing his plumage complete with black chest crescent, and the black moustache marks of the male. As soon as the young saw the returning female they began a breathy wheezing filled with urgency, but when she left after feeding them they fell silent again. This morning there is no sign of activity at the nest. The young could be crouched down, perhaps asleep, but for some reason I sense the cavity is empty. Then I hear an uncertain *queee-uw* repeated several times from a nearby tree. Before I can find the fledgling the adult male flies to the branch where it is perched and feeds it. A little distance away I hear more *queee-uw*'s, and a second adult calls briefly.

By late morning a wedge of clear blue appears on the northwest horizon. A light breeze fans the stagnant air. I set out for a small lake, the way going through poplar forest, across a meadow, and past ponds, bogs, and several fens. The track meanders and I recognize from the beginning that the lake will probably be my turning point rather than my destination.

The flower-flecked forest is dominated now by a few species: rose, purple peavine, vetch, and northern bedstraw. Twig growth has all but ceased and the tender young wood is hardening. Only occasional new leaves unfold as extension yields to maturation, to the ripening of fruits, the filling out of next year's buds, and the fattening of seeds. Winter is coming—not tomorrow, but soon enough.

At the edge of the meadow I find a cluster of nodding onions among the glaring white anemones, and dusky-pink paintbrushes. Each long, slender stem inclines under the weight of a starburst of small lavender cups, while the long slender leaves arching from the base add substance to this demurely elegant plant.

The cloud bank is gradually sliding southeast and almost half the sky is clear. I expect the sun to be unveiled at any moment, and just as I reach the far side of the meadow I am suddenly bathed in brilliant-hot light. All colours are instantaneously vibrant; highlights shimmer; cut-edged shadows intensify the brightness and themselves deepen in contrast. By changing only surface appearances the sun has transformed the landscape from suspended to flowing energy. I feel a similar change within myself, and like the plants absorb the sun's vitality.

Beside me waist-high fireweeds are kindling into bloom. The long, pyramidal spikes of hot pink flowers rise free above the attenuate, white-veined leaves. Until drained and withered by the repeated frosts of early October they will be one of the park's most commanding wildflowers. Fireweeds rarely grow singly, more usually in flocks gathered at forest edges, and they blossom for nearly two months as the flower buds open slowly from the base to the tip of the spike. While there are still flowers, even a few buds, the maroon seed pods elongate below. In early September these split open to release white-silked seeds, many of which tangle in their own threads and stay on the plant into the winter like heads of cotton candy. At the same time the leaves and stems colour, dark green being replaced by orange, vemilion, carmine, or burgundy.

Behind the fireweed a sprawling buffaloberry is heavy with ripe fruit, shiny, thin-skinned berries whose bright red is specked with white. Bears no longer come to strip the bushes, but a variety of songbirds as well as chipmunks use a portion of the bounty. The rest falls to the ground, ultimately to enrich the soil, and on the way nourishing fungi, bacteria, and insects.

Before long I find the noonday sun uncomfortably hot and walk on into the forest. For a time sounds and activity are a featureless background drone of conversation. Then I am drawn to a knot of harsh, gutteral croaks and terse caws. The three crows are in an aspen—two stubby-tailed juveniles up in the crown branches looking down at me, and a parent about half-way up the trunk on a dead branch, also directing its attention to me. I stop, but the birds are not mollified and when I pass them the young fly away clumsily, still croaking, with their parent close behind.

In time the trees ahead begin to thin and I glimpse the spruce-studded plain of a bog. Tennessee warblers have been singing persistently and now one launches from a high perch. His flight is slow and fluttery and he chatters out the second and third segments of his usual song before perching again. Several other warblers, most notably the ovenbird, waterthrush, and chat, render distinctive flight songs, but the habit is far from general, and in other families is chiefly characteristic of species who nest on prairies and tundras where elevated perches are rare.

I leave the bog and am again enclosed by forest.

A piece of dead leaf dangles in front of my eyes; at least, that is what it seems to be at first. Then it wriggles slightly and I know I have been deceived. An inch-long looper caterpillar is before me suspended by an invisible thread. It is patterned irregularly in the brown and parchment hues we associate with dead leaves. In addition, the middle third of its body is much thickened so that it appears to have the bulge as well as the taper of a curled leaf. I step back and at a distance of not more than a yard the dead leaf resemblance is perfect. The looper lengthens its silk thread, dropping several inches. It is probably ready to pupate on the ground among last year's leaves.

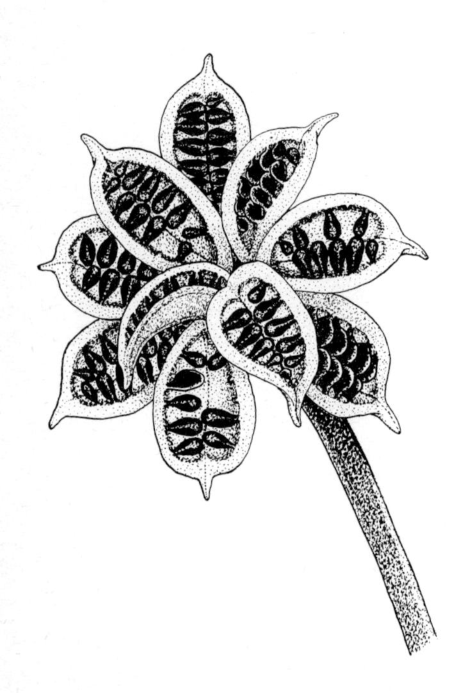

The track veers northwest and brings me close to a fen. I press through the rank grasses and herbs, and the entwined shrubs to the edge. A small brownish bird slips away silently—possibly a swamp sparrow or song sparrow. Among the sedge clumps the now huge marsh marigold leaves are spread like platters. Above them, topping each stem, are the scimitar, pale-green seed pods radiating in circles. Some of the pods have split medially and reveal neatly aligned, emerald seeds. By contrast, water arums, transitional now between flowering and fruiting, seem derelict—the immaculate hooded spathes sullied by decay; the massed yellow stamens on the clubbed flower heads shrivelled, and the young seed capsules an indeterminate green. Their glory will not be regained until the transfiguration of their ripe capsules to a deep, glistening blood red is complete. Smaller plants have grown up between the marigolds and arums—marsh nettles, their flower clusters a purple-dotted pale pink; tufted loosestrifes, their dense spheres of golden florets half hidden by the paired leaves; succulent jewelweed with only a few buds advanced to orange flowers. Wild mints, skullcaps, marsh cinquefoils, Parnassia, and western docks are all budded, but not yet blooming.

Mosquitoes soar up from the humid shade and in a few minutes the annoyance of their numbers drives me away. I leave behind a yellowthroat warbler whose reiterative singing gives the impression that he lives in a perfect world. For the present he does, and that is all his concern, not whether the present is five seconds, five hours, or five days.

Two cedar waxwings fly into a dead aspen. The female immediately begins trilling, then flutters her wings and opens her beak in supplication. The male leans over and inserts his beak in hers several times, but no food is given. Then he moves to another branch. The female does not follow him and regains her normal posture. Abruptly both depart simultaneously, flying over the trees. This token feeding between mated pairs occurs in many birds, most frequently when the union is new, and probably strengthens the bond.

I look down on the round of a little, forest-walled pond. The shore is firm and trees meet the water's edge. But from there a wide iris of cattails rings off the central pupil of open water where a flotilla of pond lilies rides at anchor. Their flowers shine like small golden bowls set on green lacquered trays. Red-wings fuss in the cattails, in and out of view, ebony males and dark-striped females. They make short flights to the forest edge, onto the lily pads, and to a fallen tree lying in the water, gathering insects then returning to nests or wheezy fledglings clinging to stems. Now and then a male takes a moment to puff his scarlet epaulettes and reel his liquid song, but most of the exchanges are whistles and *check*'s.

A new dragonfly, one of the blue darners, is on the wing, its body fully three inches long and apparently all black until I see one very close and pick out the bright blue spots and bars. Many of them cruise over the pond, while others spiral above the trees, conspicuous against the clear sky.

Recently transformed wood frogs swarm in hundreds on damp leaf mold near the water. I pass my hand low over the ground and little forms hop away in all directions. I pick one up very carefully. It barely covers my thumbnail and still possesses the stump of its tadpole tail, but otherwise is a miniature replica of the adult even to the lustrous, gold-rimmed eyes. Eventually I find a number of chorus frogs as well, incredibly even smaller.

I sit on a broad log to eat my lunch, and thus draw in my visual horizon to a few trees and the enveloping undergrowth. I am submerged, washed in a subdued aquatic light, the day's intensity far above. The air is calm, humid, and mild. Mosquitoes find me, but not in distressing hordes. A black carpenter ant walks along the log over hills of moss and ridges of lichen, down a ravine in the bark, up the mountain of my leg, and journeys on gradually angling toward the ground. A crane fly drifts past, long legs trailing helplessly. A small smooth green looper lets itself down on a thread.

A band of chickadees approaches, their vivacious chatter lifting the shrouding of languor. Soon they surround me, shaking twigs, flitting from perch to perch, hanging upside down, peering here and there and coming close to see what manner of beast I am. At first all seem to be adults, but three of the five have loose plumage, and several times when approached by a parent with food, revert to begging. The chickadees find many insects, few of which I had noticed—small caterpillars, leafhoppers, and green aphids crowding a stem tip. Although feeding continuously the birds are all the time moving from shrub to shrub, so that in two or three minutes their bright conversation fades and I am cut off from the world again.

As I sit absentmindedly I gradually become aware of a new sound —intermittent leaf rustling. I look around carefully, but can see nothing. The sounds persist. Then an animal suddenly appears at the far end of my log, a young chipmunk, dainty, and infinitely appealing. I feel a compelling urge to hold it a few moments in my cupped hands, but it is truly wild, at home where it is, untouchable. It whisks its tail and vanishes. Squeaks and twitters are followed by a rushing over the leafy ground and small plants swaying and jerking—at least two young in play. I suspect their den is tunneled under this log.

I rise, surface through the foliage, and once more take in the pond, its lilies and cattail border; and additionally now a family of buffleheads, white-cheeked female leading eight puffball ducklings mottled sepia and white. Then a bull moose raises his streaming head, grinding slowly on an uprooted lily. Both moose and beaver relish these plants, which probably accounts for their scarcity. The moose senses my presence and turns his head. His antlers are half-grown, blunted with velvet, vulnerable. However, I am not near enough to drive him from his lily feast.

The last quarter mile to the lake follows a bison trail through shrubland. The sun and heat are too much for the mosquitoes and they soon stop following. A clay-colored sparrow flies up *tick*-ing rapidly; nest or young are close, for it does not go far and continues calling. Beside the trail a mingled confusion of Canada anemones, yellow avens, rough cinquefoils, fireweeds, paintbrushes, vetches, and occasional wood lilies fill the gaps between the shrubs. Buffaloberries are red with fruit. A catbird mews hoarsely and stays out of sight. A single, quite fresh tiger swallowtail comes to rest on a wood lily. Ctenucha moths fly from shrubs and flowers along the trail. The plain, wood-brown wings do not seem to belong with the royal-blue body and the orange epaulettes and head, but when the moth is motionless the folded wings pull across a drab, concealing cloak.

As soon as I can survey the water I stop, but no birds are near this shore. I walk slowly down the bank into the shade of the old aspen. The ground is gouged and bare, while bulrushes and other shallow-water plants have been trampled down in a broad swath by bison coming to drink. But on either side, between shore and the main ranks of flower-tasseled rushes, varnished arrowhead leaves have surfaced, polished water smartweed leaves float, and burreeds disclose zigzag stems bearing solid globes of minute, greenish-white flowers.

The far side of the lake shallows more gradually. There cattails, bulrushes, and sedges congeal in broad scarves and wedges with labyrinthine waterlanes between. Ducks glide in and out of view —harlequin drakes, obscure hens, and ducklings from newly hatched to more than half-grown. Many are difficult to identify. Coots too are present, at least three families. Most of their young are well covered with a loose light-grey first plumage, but one still has some of the natal red on its head. Several pairs of black terns wheel over the water, frequently dropping down to pick up insects, then disappearing into the rushes where their nests must be. A drake ruddy duck, nuptial adornments untarnished yet, glides into the centre of a small stage curtained in green. And there, without an obvious recipient, he displays ecstatically, over and over. So regular and repetitive are the sequences that he soon seems more machine than animal, until, abruptly, the trance is broken. He relaxes, reaches forward to drink, shakes, and swims away.

My watch and the sun are at variance. The first tells me it is late afternoon, while the second seems to be motionless in the apex of the sky.

Four mourning cloak caterpillars graze on the leaves of a balsam poplar sapling. Almost ready to pupate, they are about two inches long, red spotted on a black background, and armed with rows of formidable black spikes.

The rest of the walk is uneventful largely because I am dulled by the heat and more conscious of my own day-end discomfort than my surroundings. A robin flies and pauses on a branch. Pale-bordered belly feathers and a few spotted ones on the upper breast denote its immaturity. At the meadow bison have just gone, leaving insects the only animal claimants—dragonflies, damselflies, bumblebees, hover flies, orange and sepia fritillaries and pearl crescents, yellow alfalfas. I start up one tailed-blue from the grass, and a brown and buff wood nymph.

Late in the evening, just before sunset, I sit in the cool by the lake. Coyotes yelp briefly from the north shore; a sora rail whinnies; gulls assemble for the night, while high overhead a feeding nighthawk jinks, and calls, and booms out its parabolic dives.

I make a very early start and reach the Sandhills east of Moss Lake by eight o'clock. The sun is already high, but in shade close to the ground plants are still soaked with dew. The land here is only slightly rolling, like gentle ocean swells, and neither meadow nor forest. Scattered groves and individual aspen rise out of low shrub thickets, in places too dense to walk through, elsewhere with streams and pools of grass between. And spread through the grass, for nowhere does it form a compact turf, are islands of pussy-toes and white-capped yarrow, clumps of mauve fleabanes, and pale violet harebells together with stands of paintbrush and wood lilies. Slender blue beardtongue, although common, is not a showy plant, growing little more than a foot high and producing small, dark flowers. Ripening strawberries glint bright red under their canopies of leaves. Bearberries spill out from beneath the taller shrubs, layering reptilian-scale leaves over the bare sand.

A troop of magpies streams past laughing, flapping loosely, long tails floating. They all stop briefly in an aspen before continuing their journey. Magpies give the impression that sustained flight is an effort for them, even in calm.

Tan bodies are suddenly bright against the green wall of vegetation. Then three cow elk slip between the trunks and almost immediately disappear. I think I may have seen the smaller form of a calf, but cannot be certain.

Clay-colored and savannah sparrows sing from shrub-top perches; drop down and call warnings when I pass near. A pair of goldfinches is in the air and receding, rolling over unseen waves, whistled notes melodiously mimicking their flight. Even later to nest than cedar waxwings, goldfinches await the ripening of cotton-tufted flower seeds, both food and nesting material. Often the young do not fledge until late August.

By mid-morning the sun is warm and little cumulus clouds flock the sky, too small and widely spaced to cast shadows. Against this blue and white a red-tailed hawk soars effortlessly in wide circles. But instead of mounting endlessly higher it slides off the thermal and drifts sideways.

I begin the walk back, by pale-flowered buckbrush humming now with bees, hover flies, and wasps, and fluttering with butterflies. Grasshoppers shrill from many points, in crescendo when several coincide, but more often in thin isolated foci. Orb spiders have set their webbed traps in the shrub thickets and in one, torn badly by the struggle, a swallowtail hangs dead and trussed. Amber and scarlet skimmer dragonflies dart about and occasionally pause on leaves. The large darners hawk constantly, sometimes at eye-level but more often near the tree tops and above. At the edge of a moist depression tall grasses nearly hide a patch of plains arnica, sturdy plants displaying big, yellow-rayed sunflower blossoms.

I stop at the kestrel slope. The female is on the topmost branch of the nest tree twisting her head from side to side and upwards, and apparently following the darners. Suddenly she flies in a direct line using shallow rapid wingbeats and close to one of the insects swerves abruptly coming at it from slightly below. But it dodges and she misses. She returns to her perch at once. There she remains for at least five minutes, occasionally preening quickly, but still watching the dragonflies. Again she flies, as before, and this time is successful with her forward reaching claws. Back on her perch she deftly nips off the wings which descend spirally, flashing in the sun. The rest of the dragonfly is eaten in a few quick bites. The kestrel cleans her beak on the branch and resumes her watch. Twice more in the next quarter hour she catches and eats two more dragonflies in as many attempts.

Nearly as large as a crow, but slate-grey above, a Cooper's hawk angles across the slope carrying a bird or a mouse in its claws. At the same moment I see it the male kestrel appears and dives at the larger hawk with piercing cries. It ducks from each plunge and on reaching the trees immediately drops out of sight. The kestrel does not pursue it farther, but sheers off and perhaps returns to his lookout perch. The female has taken no part in this episode and when I leave the slope she is still at the top of the tree engrossed with dragonflies.

It is a long hike to the remote bog where sundews grow, but although today is sun-flooded the early afternoon is no more than pleasantly warm. Nearly all the distance is through dense mature poplar forest with only a few fens and ponds to interrupt its continuity.

At a pond cattails are flowering, smooth brown-green cylinders heading slender shafts. A constriction above the mid-point separates female from male flowers, and since they are wind-pollinated, structure has been reduced to simple stamens and pistils without any adornment. Nearer shore purple-blossomed marsh cinquefoils retire diffidently, have to be sought out in the sedges, but not the soaring white domes of water parsnip, although these are smaller, more refined plants than the similar cow parsnip.

Red-wings are broadcast through the cattails like dark seeds—males, females, and many short-tailed, inept, pleading fledglings. A small bird flies in over the trees, drops lightly onto a half-submerged log, and begins walking along tipping its tail end up and down, dipping its beak into the water. It is a spotted sandpiper, but a juvenile bird with clear white underparts and clay-brown above.

A complete family of ruddy ducks slips out into the open. Unlike most other ducks, the drake assists with rearing the pair's offspring. The ten young ducklings follow the hen in a body, while the drake shepherds the whole group. He displays several times, flipping up his fanned tail and going through his jackhammer jerking. Still in nuptial dress he exhibits another of the stiff-tailed ducks' differences. Although they moult twice a year, the drakes do not begin to develop a cryptic appearance until well into August, nor return to breeding plumage until the following April.

The trail rises and falls, rises and falls, curves around wet hollows and occasional bogs; seems unending. Sometimes I feel I am not making any progress until I pass a plant I know I have not seen before. One time it is a few stems of spotted coral root, the whole of this saprophytic orchid in dull reds; another, a group of baneberries with the first ripe fruit, big china-white berries. In between, and repeating like a patterned fabric, are yellow agrimonies and fringed loosestrifes, purple peavines, pale-flesh dogbanes, and vivid pink fireweeds. Currant and gooseberry fruits are ripening, as are a few dewberry and the occasional raspberry. Here and there dogwoods have put out new shoots in compensation for ones browsed, and fresh white flower clusters gleam beside half-grown, leaf-green berries.

A bird flies across the track; perches. It is a young blue jay, full grown but loose-feathered and its blues duller than those of the parent who follows it, then flaps on through the trees calling.

At the edge of sedge-thick hollows a few Parnassia spread solitary white flowers below the taller, lax heads of Philadelphia fleabanes, but do not intrude on the closed formations of orange-bannered jewelweeds. And in these damp places I again encounter recently transformed wood frogs and toads. Most are now tailless, yet still cover little more than my thumbnail.

I turn off the trail to reach the bog, and in a few minutes dark spruce spike above the poplar crowns. Then I am there, with only a slight slope and an abrupt change in the vegetation. The spruce are like a mountain range guarding the chastity of a lush, unspoiled valley. But by bending low, sometimes to my knees, I find a way through. The trunks are stepped into a dry deck of brown needles that lighten the gloom hovering in the branches above. Sun rarely penetrates, but where it does it has raised small, velvet moss cushions. The air is dry, slightly musty, and laced with resin.

I come to a small opening. Simultaneously a Cooper's hawk blunders off with a harsh, prolonged rattle; yet not far, and continuing its rattling call. Ahead an old stump is spattered with white droppings and its rounded top is plastered with dried blood, broken feathers and wisps of down. Remnants from many meals litter the soft green moss below—a cedar waxwing's red-tipped wing feathers are the most conspicuous and recent. It must require many birds and small mammals to sustain a pair of Cooper's hawks with their nest of four or five rapidly growing young, yet they range widely, taking one here, another there. Songbirds in the vicinity of the bog do not seem unusually scarce.

The gloom recedes, then I leave it behind as the spruce shrink to saplings. Cotton grass is in seed, tufting the squashy inner bog with white. At any other time the plant is just another sedge, thin-leaved and anonymous. The bog centre is closed over by a thin skin of Sphagnum mosses, but long before I near it the raft of tussocked sedges, bog rosemary, blueberry, and cranberry quakes. At my feet I find a few sundews, tiny in spite of their predatory habits, and able to hold nothing larger than a house fly. The leaves, like long-stemmed spoons, radiate out so that each blade has its own space. A multitude of sticky-headed red filaments reaches up from the flat blade, waiting. An alighting insect first finds its feet stuck; then adjacent filaments curve over gluing the wings, legs and body; finally the whole leaf folds. Once the valuable proteins have been assimilated, the plant's chief source of nitrogen, the leaf will open again and in time rain will wash away the indigestible husk. A slender flower stem ascends from the centre of one sundew, with a few miniscule white flowers near the top. The plant itself is perennial, and its blooming seems incidental.

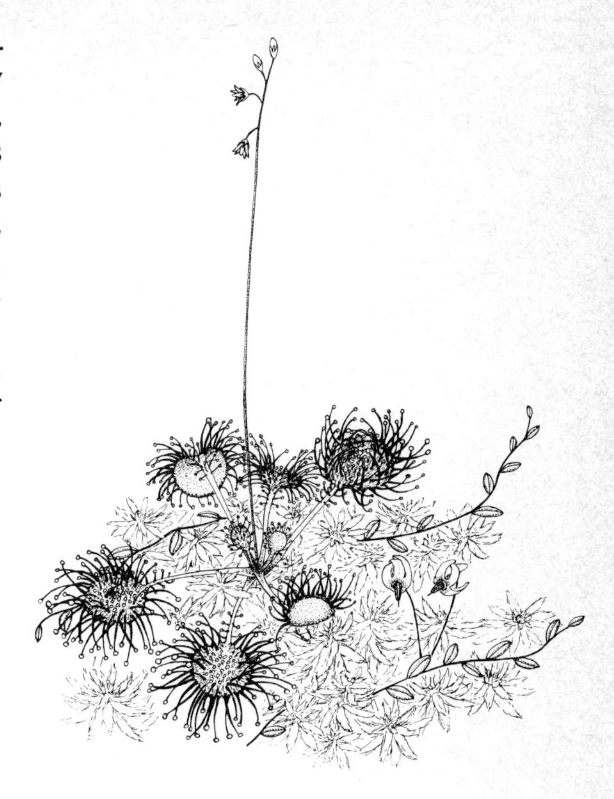

I depart wondering how many moose skeletons lie buried in the bog's centre. It must be a death trap when ice first crusts the surface.

By late afternoon I am at the flicker nest, but find no sign of the family. It is now a week since the young fledged.

On the way home I notice many balsam poplar leaves with a large globular gall at the base of the blade. Finally I open one carefully. It is hollow inside, like a gourd, and packed with many aphids, most wingless and pale green. The occasional winged ones are hoary grey. I open three more galls, two with a small hole through the wall. These latter house fewer aphids of which perhaps a quarter are winged.

Early evening beside the lake. Pink, club-headed smartweed flowers among white, three-lobed arrowheads lacing the shallows, rocked by wavelets. Countering ripples lining back from the head of a swimming beaver out beyond the edge of the pondweed. Ducks clotting the water like wind-drifted flotsam.

A great blue heron approaches from the north shore. It veers toward me, slows, backstrokes, and drops, its legs stretching for the water. Then it is standing folding its huge pinions carefully as though haste might tangle them. Once I saw it intended to land I drew back deeper into cover hoping its glaring, yellow-eyed stare would pass over me unseeing. I think now it has, although the bird seems tense and suspicious and looks back to the land rather than into the water. But gradually it relaxes. So near, less than thirty feet, it is more than ever an imposing, elegant bird, spare and angular yet well balanced. It folds its neck down between its shoulders, only the head and beak leaning forward, and remains completely motionless for several minutes, waiting for suitable prey to swim within striking distance. None does and the heron begins to walk very slowly parallel to the shore. Now its neck is slightly extended as it stares down, absorbed in its quest. A spearing jab—a stickleback crossways in the beak and quick, gulping swallows. The heron wades on and makes several more strikes before a bend takes it out of sight.

I ease out of hiding. This day never lost its fresh morning clarity and the evening is crisp and vitally alive. The setting of the sun equals its resplendent rise, burning on broken streamers of cloud. Gulls fly inbound on the same courses they traced out to their day's foraging. The songs of birds falter and cease in the reverse sequence of their chiming in. The daily cycle has swung about the pivot of the sun. The heron reappears flying ponderously south, the last slanting rays lightening its sombre hues.

Hot, heavy, humid air billows up from the Gulf of Mexico, comber upon comber of an invisible ocean pushing aside the clear northern air. The sky's brilliant blue fades and is weighted with haze. The unshaded sun is a blade of fire thoughout its interminable day, pushing the thermometer into the mid-nineties even in shade. Nights are too short for deep cooling, or dew. The dearth of clouds, save for a few seed cumulus, make the open intolerable; swarms of demented mosquitoes, finding the trails where my running sweat has washed away the repellent, make the forest intolerable. It is too hot to work, to eat, to sleep. The lake is tepid, crawling with life, green-skimmed with algae.

Birds stop singing. Bison must graze only at night for all day they are to be found in shade, lying or standing listlessly, tails switching ceaselessly, feet stamping, ears wagging. Deer, elk, and moose seem to have vanished. One night, dark and moonless, I hear coyotes briefly, the only occasion in five days.

A wind stirred last night, restlessly snuffling through the leaves. This morning cirrus streamers trail across the sky from the west. Stratus thickens behind; cumulus build below. Through the afternoon the lower clouds coalesce, gaining towering height and burgeoning width. The sun is masked, but the heat does not lessen, if anything is more oppressive imprisoned under the sagging cloud ceiling. Like an army massing for an assault the storm line deepens and darkens on the horizon. Its advance seems slow for a long time, but as it nears its corpulent front comes rushing towards me. The light dims and the last birds leave the sky. Incongruously, at ground level the air hangs breathless.

Lightning flickers; twenty seconds later a gutteral growl. The clouds twist onward like flowing lava and almost as dark. The lake lies flat, an expanse of oily pitch. Dusk is real and all animal sounds quieten. Still the air holds its breath, awaiting Armageddon. Lightning flares here and there every few seconds, sometimes in narrow crooked threads, sometimes in broad sheets of reflected light. The growls become roars interspersed with staccato reports.

Wind grabs at the tree tops, passes through them and is gone; then a second lash from a different direction. Suddenly cold blasts down and the wind, a river in spate, tears through the trees to the ground, rears waves on the lake. Rain plunges in big, heavy drops. It is almost dark. The sun must have set, and for a few moments I sense its separation from earthly events, the prime mover unmoved.

Soon lightning and thunder fuse in an unrelenting barrage—electric-blue light startles the trees, sears my eyes; the air explodes and crashes, reverberates and echoes. Drenching rain pummels the earth and blanks out all the world beyond the garden trees; changes abruptly to white hail bouncing elastically back from the ground, colliding, falling, bouncing, subsiding, covering the ground. For ten minutes hail streams down battering, ripping, blanketing. With every strike light reflecting from the whitened ground floods through the room. I begin to feel vulnerable, as if there were nothing more between me and the elements than a transparent plastic bubble. The hail hammers on the roof, the walls, the windows; curtains stir uneasily in wind-driven draughts; each flash engulfs me in blinding luminance.

I cannot see again for several seconds; my ears are split by air rent asunder. The house shudders as the anguished air self-destructs in cacophony. The next glare, farther away, reveals the damage—a tall old poplar near the shore with the top third of its steaming trunk cleft, and one half leaning over, branches tangled in those below. The hail has reverted to rain, although the wind thrashes on undiminished, trees and shrubs leaning away from it, sacrificing leaves, twigs, and occasionally whole limbs.

Over the next half hour the violence slowly subsides as the storm centre tracks eastward. The lightning is less frequent and dims; the thunder retreats. I can hear the wind now as a separate entity, but its force too is nearly spent; likewise the rain which has settled to a steady falling.

All danger past I go out briefly, scuffing through the hail, squishing in the sodden earth. Streams run gurgling along the hard-packed track. The forest itself is a dripping cavern, a lost, subterranean realm. The air feels very cold, saturated, a river flowing through my lungs. At the edge of the lake invisible waves rolling across the wide fetch of the bay smash themselves against the end of the point.

Later my sleep is broken by other storms embedded in the cloud pack, but none approach the stature of the first.

Days after the storm its trail is still obvious. Fragmented leaves curl dry in the sun, the canopy above noticeably thinned. Leaves not torn off are scarred with brown-edged tears and segments ripped out. Grasses and herbs are broken, many shredded beyond recognition; fruits are gone from shrubs; tight-lipped green cones are scattered under the spruce; a waxwing's nest lies on the ground, its dead bedraggled nestlings drawing flies and beetles. But the swath of destruction is not much more than a mile in width, as if some giant, dull-bladed harvester had churned through.

❦ ❦ ❦ ❦ ❦ ❦ ❦

Showers pattered again last night, but a stiff, northwest breeze begins to break up the cloud early and by noon it has all blown away. I take a lunch and spend most of the day near a large beaver pond that dams up water at the head of a chain of wet sedge flats and small pothole ponds. Before the founding pair of beavers established themselves here a few years ago the pond cannot have been more than half its present size. Lines of drowned poplar mark former ridges extending close to the centre, while old stubs, skeleton willows, and a few leaning trunks rise out of the water some yards from shore. During their tenancy the beaver have felled many tall aspen. The occasional one is still hung up in the branches of a neighbour, but the majority lie prostrate or angle into the water. Most of the branches have been cut off and large patches of bark gnawed away. Of the mature trees that still stand close to the water, almost all are balsam poplar. The beavers must realize that an aspen yields more food for the effort of cutting it down since the bark thickens only on the basal two feet or so of the trunk. Young aspen and balsam poplar, however, seem to be cut with equal frequency.

Two adult kingbirds perch in drowned trees not far apart. They call frequently, perhaps to each other, for the notes have less of an edge of annoyance than usual. From time to time they sally out after insects and I hear a click each time the near bird snaps at its prey.

Towards the far side of the pond a pair of red-necked grebes stays by a grey-fingered willow, swimming slowly and preening, but never moving far from where I first see them. Perhaps they are still nest bound. Although these grebes begin nesting early in May, loss of either eggs or very young chicks will stimulate relaying as late as the end of June.

The only other waterbird I locate is a female bufflehead resting on a half-sunken log with two downy young beside her. These too are late, and probably from a second clutch.

In the water are many dark green leeches, palely spotted on

the upper surface. Some are arched longitudinally and when one tilts sideways slightly I see a crowd of young leeches clinging to its underside—its own progeny which it is protecting, although not feeding.

I tread the narrow margin between the wet sedge fen, fringed with willows and alders, and the rising slope thick with mature poplar. A yellow warbler sings intermittently from across the meadow, at least I hear it only when a yellowthroat a few feet away ceases singing to pick up an insect or change his position. Yellowthroats always seem ready to investigate any novelty in their territories and often come up to a human intruder without encouragement, although not showing the confiding curiosity of chickadees and kinglets. Nevertheless, the warbler follows me as I walk along slowly.

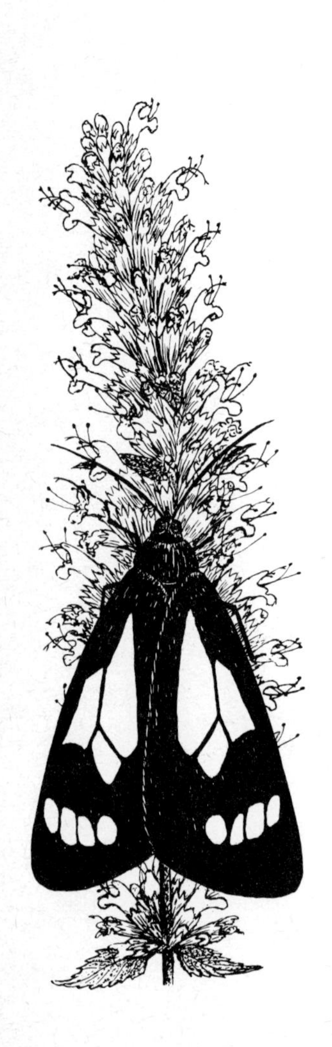

Wild mints and skullcaps bloom in purple and blue generosity under the willows, flanked by wisps of sweet-scented bedstraw and clusters of Parnassia wherever the grasses and sedges thin. At the edge of the forest black twinberries are ripe. The long yellow-belled flowers of late May have been replaced by pairs of polished black spheres pendant from a parasol of red-flaring bracts. Farther back purple-spiked giant hyssop spire above the welter of the undergrowth. As a beacon light in the dark they have attracted broad-winged pericopid moths, on one plant numerous enough to mask the flowers completely. The moths are about an inch and a half long, wedge-shaped at rest, velvet black and flash white. Additionally, a bib of long orange hairs hangs below the mouth and covers the femur of each front leg. Today these moths are suddenly abundant, but I find them almost exclusively on hyssops, fireweeds, and dogbanes, or dithering with rapid wingbeats from one plant to the next. One mating pair on a dock leaf are joined, abdomen tip to tip, each the mirror of the other. White admirals are also plentiful, sometimes at the same flowers—black and white moths, black and white butterflies.

I am nagged by a family of least flycatchers, two adults and three young. The fledglings maintain a rarely interrupted stream of dry *chip* notes which changes quickly to the universal wheezing when a parent brings food. From time to time the male delivers his *che-bec*'s and once gives chase to another adult, presumably trespassing, for shortly after they separate both the close and a more distant male commence hiccuping responsively.

The fen widens and I find myself following a bent-sedge trail into the open. After about twenty yards it terminates at a rough oval of flattened sedges where a moose bedded down. When I lie too, there is room to spare at my head and feet. Overhead the clear sky, a deep singing blue, begins at the sedge tops and extends infinitely far, or no distance at all. A fritillary passes above, yet does not change the relationship of the sky. The sun beats down into my enclosure and soon I begin to feel hot, stifled, and slightly claustrophobic. Standing again I can see where the moose continued on across the flat. The breeze has strengthened and rolls unhindered across the sea of green in glistening swells.

I cross where the fen narrows—balancing precariously to step over the entrenched seep from the beaver pond and squelching a few more feet to gain solid footing. I strike an old bison trail and follow it for some distance over low, forested ridges. A pewee slurs plaintively above, and a male oriole chatters loudly in flight. As I walk I pass from one red-eyed vireo's territory to the next's. An occasional robin clucks like a hen. In a place of stumps and old blowdown, two house wrens stutter excitedly and the male sings once. I pause and soon find the reason for their agitation—three young fledglings lined up on a dead branch, very still, waiting for the danger to pass. An immature sapsucker, brown where the adults are black, and lacking the distinctive head pattern, hitches up the bole of an aspen. Every few seconds it pauses and appears to inspect the bark, but does not attempt to drill any holes. Then it flies, giving the species' typical hoarse, lurching calls. A Compton tortoiseshell detaches from the path ahead, foreshadowing autumn in its wings.

There are few fresh flowers now in the heart of the forest—hyssops, dogbanes, fringed loosestrifes, agrimonies, purple peavines, a scattering of roses, white wintergreens, and northern bedstraws. Ripe fruits are taking the place of blossoms—deep red strawberries, dewberries, and buffaloberries; shining white or red globed baneberries; purplish-black beaded sarsaparillas; and grape-hued gooseberry spheres. The hazels' green velour husks are nearly mature and swollen at the end by the nut.

I lunch looking down on a willow-studded fen. The heavy blurred heads of wool-grass nod stiffly in the gusty breeze, in contrast to other sedges that yield unresistingly to its pressure. Somewhere out in that anonymous expanse a LeConte's sparrow buzzes faintly, so near to the rustling of the leaf blades that I wonder if my imagination has created him for my ears to hear. The swamp sparrow who flies to the top of a willow, sings once, then flits off to another willow, is scarcely louder. But perhaps the sparrows are the catalyst, for a yellowthroat warbler suddenly begins his hurried *witchety-witchety-witchety-witch*, only to be overcome a few moments later by a Connecticut warbler. Twigs vibrate, but I never see the birds themselves; and in a short time the singing falters, then ceases altogether.

Soon I hear chickadees approaching from both sides. They resolve into a family of six, and one of four. They appear to be busy feeding, but as the distance between them narrows their calling intensifies—a constant patter of slurred twitters punctuated with single *dee*'s, a few *chick-a-dee-dee-dee*'s, and once the bright whistled song. But in the end each group remains cohesive, no chases occur, and they maintain a gap of about twenty feet. After two minutes or so in such close proximity the two bands gradually drift away from their territorial borderland and the calling quickly subsides.

The bison trail crosses an overgrown bulldozed track and I turn onto the sun-filled defile. A jay screams once, like a bagpipe skirl, and planes across the gap in a flash of blue. Outside the breeding season this bird is garrulous and brashly inquisitive, so seems more numerous than it actually is. But from the moment the nest is begun until the young are well fledged it is silent and secretive, so seldom observed that it might have migrated.

Forest, edge, meadow, and "weed" flowers comingle in a kaleidoscope of colour along the track. Spike poplar and willow shoots will be crowded saplings a decade from now as irresistibly the forest heals its wounds. Hover flies, bumblebees, solitary bees, yellow jackets, hornets, flower beetles, white admirals, pearl crescents, fritillaries, alfalfas, Ctenucha moths, broad-winged pericopid moths, and a hummingbird moth are all visitors to the flowers. Froghoppers and leafhoppers spurt from plants and the grass at my feet like miniature grasshoppers, although they are most closely related to aphids. Without a hand lens or microscope adult froghoppers and leafhoppers are difficult to separate, but the nymphs are at once distinct, for the former are the familiar spittle bugs whose bubbly white foam masses are conspicuous on many herbaceous plants. The nymphs of the latter, although they also imbibe plant fluids, do not produce foam. Damselflies, the majority blue and black, and skimmer dragonflies, some amber, some scarlet, patrol the air, occasionally dropping down to a leaf to rest or groom. Insects accomplish this with the legs alone, sweeping them over the body, including the eyes, so that the hairs and spines pick up dust. To conclude, the legs are rubbed together much as we would clean one brush against another.

A shrubby clearing opens out to one side, but appears to have no natural cause—the slope is gentle and faces northwest. Perhaps this is the site of an old homestead cabin and garden plot. Whatever it once was, it is now a jungle of interlacing, springy stems and spindly poplar saplings, dappled with the pink of fireweed and paintbrush. Budded goldenrod and asters are near to opening.

As saskatoons ripen the berries turn from dull red to a rich, shining purple and, at the peak of readiness are glazed with a thin, whitish bloom. Unwilling to wait for perfection, waxwings and robins are already plucking the dark fruit. A few pin cherries are tinting red, but choke cherries are like hard, spherical emeralds, except on one bush where the developing fruits are distended and distorted. When I open one of these I find a cavity filled with vivid pinkish-orange, squirming gall-midge larvae. I suppose I have seen the adult flies, but without recognizing them as anything other than small "midges."

A sun-spangled orb web bars my way and as I look for its builder an inchneumon fly blunders in and is held. Its thrashing soon brings out a dark spider, considerably smaller than the half-inch ichneumon. The spider inspects its catch briefly then, without attacking, runs lightly back over its tightrope to its retreat in a curled, silk-bound leaf. Perhaps the prey is too large, or perhaps it is because it is a hymenopteron, a group many spiders reject regardless of size. The ichneumon continues struggling, jerking and swaying the entire web, until it pulls free and flies off swiftly.

Before I leave the clearing a chipmunk stutters its monotonous *chupp*-ing alarm, telling all that a stranger is at hand.

Small, metallic-green dolichopodid flies dangling long slender legs trail back and forth over a large pool in a hollow in the track. Frequently some pause on the surface in strange, uptilted stance. They are probably looking for helpless insects their own size or smaller. This family includes active predators in addition to scavengers, although the predators are cursorial rather than aerial like robber flies, hence the long legs.

The little bog, set deeply down, is so reserved, so much a keeper of hidden truths and myths that I go no farther than the outer edge of the white spruce barricading the rim of the depression. No sound at all comes from it, although a squirrel scratches on bark as it moves about the white spruce.

By late afternoon I am back at the beaver pond. Already the dam and lodge lie in shade and I sit by an alder close to the water. The shrub brings together three seasons and two years—occasional fresh leaves are unfolding, yet at the base of mature ones next year's buds are large; old, blackened seed cones hang in clots, forgotten, while this year's are mature but green, and next year's male flower catkins are half-grown. In little more than a month, however, shortening days, then frosts will telescope the stages and fix the entire plant in dormancy.

Two tree swallows curve and recurve over the water, as often across the breeze as into it; now high, now swooping low, now skimming the surface and drinking. I search for ducks, but the drakes have left to wing-moult on a lake and most hens and young must be resting. Two cedar waxwings fly into a dead tree—a pair? Then I am no longer in doubt. The male, stiffly erect, beak uptilted, crest flattened, wings held close but tail slightly fanned and pumping slowly up and down, hops sideways along the twig, three steps toward then three steps away from his mate. Brief pauses separate these dance movements, but in them the male's posture does not change. The display continues for three minutes and might have lasted longer, but a third waxwing arrives and perches near the couple. The male stops abruptly and almost at once the three birds fly away together trilling.

A fly alights on my hand, about the size of a house fly, but very slim and tan in colour. The long wings fold neatly over the back. Soon it heads out over the pond close to the water, for it is a sciomyzid, or "snail fly," and its larvae prey on or parasitize snails which here are nearly all aquatic.

A beaver swims slowly toward the dam—with only a sliver of the head exposed it still has full use of all its senses. In front of the spillway it pauses for several seconds, then veers off to the far shore. The recent heavy rains have increased the overflow and washed out some of the fill. Soon the beaver returns trailing a leafy branch clamped between its teeth, swims directly to the weakened area and jams the end of the branch into the dam. Using its jaws alone, it continues pushing the branch until it will go no farther; then with its forefeet changes the angle slightly. Another, smaller beaver swims up and together the two spend a quarter-hour bringing small branches, twigs, and muddy lumps of sedge which they pack into the dam. The overflow narrows and diminishes, but the spillway is not completely blocked. The young beaver dives and I do not see it again, but the adult comes to the bank where a patch of horsetail blankets the mud. Out of the water it is immediately cumbersome, shuffling arthritically, its spatulate, scaly tail dragging lifelessly. The beaver does not walk far, but begins grazing the horsetails, tearing and crunching them audibly. It eats hastily, at times using its forefeet to pull more stems to its mouth. Then something alarms the beaver. With incredible speed it swivels around, slides into the water, and slaps down its tail with the crack of a rifle shot as it dives. Only the expanding ring of ripples recalls its presence.

The breeze slackens. At the kestrel slope I find the whole family in the nest tree, the parents augmented by three fluffy juveniles. Two are females, for unlike most birds the first plumage is similar to that of the corresponding adult. One of the fledglings is picking at a dragonfly, and as I wonder how it could already be skillful enough to catch such dexterous prey the adult female wings out, snatches a dragonfly from the air, and returns to the tree to give it immediately to one of the young birds who encourages her with vibrating wings and shrill, wheezy calls. Now the adult male flies and on his first attempt secures his prey; but instead of returning to the tree he continues flying across the slope drawing his taloned foot forward and bending his head down to meet it. One by one the insect's wings are pulled off and skid away. Two minutes later the male reappears, and this time gives his dragonfly to a fledgling.

After dinner I walk along the west shore of the lake. I am deep in cool, pelucid shadow, but the water and islands beyond are still flooded with bright dayshine. Close to Long Island fifteen red-necked grebes and a dozen horned grebes mingle in a strung-out flock. All seem to be adults, but at this distance I could easily pass over the grey-brown juveniles. The red-necks occasionally dive, or swim slowly, peering with their heads just below the surface, or preen. The horned grebes either float passively or preen. Remaining a little apart from the other grebes are two pairs of western grebes, conspicuously larger than their relatives and with black-crested heads carried aloofly high on slim, beacon-white necks. I watch these unusual visitors for many minutes. For much of the time they glide slowly or preen; but now and then two swim side by side for several seconds, necks stiffly erect, beaks and heads horizontal, or alternatively, separate then turn and, with the same posture, swim to meet face on.

A red-tailed hawk drifts into view, circling but not really soaring. It silently trends southeast over the water and is soon swallowed up by the island's forest. A little later two magpies beat over to the island, coming to rest in the top of a tall spruce on the shore.

When I check the gulls and terns carefully with binoculars I find several immature Franklin's—near white below, but with the top of the head, back and wings mud-brown. In flight a broad black band terminates the white tail. All the common terns are adults.

The cattails in flower are like an army of lances, each tipped with gold-dusted, bronze-green points. Ducks slowly swim out from cover. Some disturbed by my walking move away quickly, looking back repeatedly yet not sufficiently alarmed to fly. I stop; soon the birds calm and begin feeding. A redhead and her three downy, mottled young gather in a knot and make short, frequent dives, although all are never underwater together. When they surface the ducklings pop up as buoyantly as corks. A mallard leads seven ducklings who are less than two weeks old. In the same area I count twenty-six juvenile widgeons, mallards, and teal, nearly full grown and probably almost ready to fly. Around the clump of dead willows "eclipse" drakes of several species form a ragged flock of about fifty. Peripherally several ring-billed gulls sit so lightly on the water that they seem to be floating above it. Common terns whiten the willows and fly among the Franklin's gulls. Near a beaver lodge built against the shore three young pied-billed grebes dive for food. They appear full-grown and independent, but retain the immature features of a plain horn-coloured beak together with brown and white striped cheeks and upper neck.

Most diurnal insects have settled for the night, although a few lacewings flutter away when I push through some shrubs, many dragonflies quarter the sunlit heights, and mosquitoes gather strength.

The sun seems to have been setting for a long time, as though reluctant for the moment of leave-taking. Now its haze-filtered light is lurid orange on the island's trees. On my way home in the afterglow a few rain-revived chorus frogs produce reedy creakings, a coyote family yelps, and a white-throated sparrow flutes a single refrain. In the rising dusk half an hour later a distant veery whispers eventide.

Three days before the end of July a companion and I set out for the Soapholes in the early morning cool. Grasses, flowers, and low shrubs are hoary with dew, spider webs beaded and sagging. The sun diffuses softly through the trees. A Tennessee warbler fanfares in the day. We cross a small meadow and our feet are soaked. But others have been ahead of us, bison dragging verdant trails like the ones we leave behind. Then we are in the forest where the dew is less and the air pleasantly mild, leaf-scented and still.

The track winds without climbing. We pass neither ponds nor bogs, only the occasional sedge flat and alder-thick hollow. The forest is uniform poplar in every direction as far as we can see. Variety is generated by minor irregularities and animal encounters. A damp depression is covered with jewelweeds whose bright orange flowers remind me of snapdragons, although they are not arranged in a dense spike, but spring out individually from the upper leaves on slender, arching pedicels. Closer inspection reveals that the three petals surrounding the open mouth are separate from the spurred funnel behind, that funnel being a highly modified sepal. Botanists believe that all flower parts evolved from bract-like leaves, so it is not surprising that sepals can sometimes look like petals in form, texture and colour. The six-parted flowers of lilies, iris, and arums consist of three petals and three sepals, while anemone, marsh marigold, and bunchberry flowers, only three examples, are entirely sepals. I touch a couple of the larger green seed pods on one jewelweed and they justify its alternative common name, touch-me-not, by snapping open and scattering the seeds. Without my interference the pod would not have ruptured until the seeds were fully ripe.

The human presence is too much for a big toad who hops away from under the plants to find a more remote hermitage.

Around a corner and we are suddenly upon a family of ruffed grouse. Seven rocket up from the ground, wings hammering, knocking into shrubs and branches. But they only scatter into the nearby aspen crowns, then look back to find what caused their fright. I locate the adult hen, crest feathers raised, clucking rapidly, and neck stretching thin as she peers about. Eventually we find three of the others, all juveniles about two-thirds grown. As long as we are still the grouse remain in the trees, but as we walk on they all fly again noisily in disorderly retreat. Red-eyed vireos sing on without interruption. A robin flies up from the track, calling, and alights on a branch. It is an adult male, but his plumage shows the wear of the nesting season, a general air of roughness in addition to frayed tail feathers.

A cavity at the base of a young tree is half-filled by a yellow-jacket nest. A single round hole in the banded wasp-paper roof provides access, and bending close I can see the faces of four guards just within the entrance. Workers continually enter and leave, those waiting to go in walking about on the surface of the nest. Up to a hundred feet away these little wasps are very common at flowers, especially fireweed, dogbane, agrimony, and the first asters.

A female downy woodpecker sits closely along a leaning poplar trunk, jerks her head from side to side several times, then flicks out her wings, but in a special way. First the right wing, three times in succession, is snapped up vertically above the body where it forms a narrow triangle. It is held there for a few seconds, then is snapped back to the side. After a brief pause the movement is repeated. Then the left wing is put through the same motions, only now I see the flash of the pale wing linings. For perhaps half a minute longer she is motionless, then flies rapidly and directly out of sight. I am witnessing a recrudescence of sexual display, although no other downy woodpeckers are near.

This track is a bison throughway. Their hooves keep a wide surface of bare earth churned up and droppings are everywhere. Time and again bevies of rove beetles scuttle from old drying pats. Coloured like the dung, the beetles are invisible until they sidle away with long, tapering abdomens curled at the yellow tip like miniature scorpions. The threat, however, is all bluff.

Something rustles in the leaves, and stepping off the track I ferret out a white-throated sparrow who has been feeding on the ground. It flies into a shrub with a thin *tink*, and for a few moments I see it clearly. Although it is in streaked juvenile plumage, the pale throat patch and broad light line over the eye are distinctive. The bird drops back to the ground and resumes feeding.

By mid-morning we come to the first of the upland meadows—narrow, forest squeezing in on both sides, trail running down the centre. The light suddenly seems too bright, too hot; colours are garish, shadows uncompromising, like a flat-toned poster beside a richly and subtly tinted masterpiece. As my senses adjust, however, the meadow becomes more acceptable. The first thistles, smooth asters, meadow fleabanes, wild lettuce, goldenrods, Canada hawkweeds, and sow thistles bring pink, mauve, and lilac in apposition to deep cadmium yellow, colour complements that vibrate with an intensity neither alone would possess. Yarrows float among them like puffy white clouds.

Several red-eyed vireos move restlessly in the saskatoons picking the ripened berries. Once an adult feeds a green caterpillar to a juvenile. A male sings twice, but is more interested in his berrying. Otherwise the birds are silent until we move closer and elicit a scattering of petulant *wheuw*'s.

Without warning we are engulfed by loud, porcine grunting. It is like a sudden rapids in a river, and flows along the far side of the meadow. But we can see nothing of the bison herd back in the trees. We slip through the outgrowth until we are among thick aspen trunks, yet still able to survey the open. For a few minutes the only change is a slight decline in the grunting as the bison move farther away. We are about to return to the trail when a massive bull steps through the trees at the far end of the meadow, perhaps a hundred yards away. He is like a boxer entering the ring—great woolly head lowered, sun-glisten highlighting the curves of his black horns and the bunched muscles of his lean hindquarters. He grunts several times, paws the ground, shakes his head. For a little while nothing else happens. Then a second bull of equal size emerges from the trees. He too grunts, and paws, and shakes his head; then goes down on his knees in a half wallow. When he rises again the two face off, crook their short tails, gather—in a short rush clash head to head. They back off a few steps, then stand, heads and tails low. I remember my camera and screw on the telephoto lens with less than steady hands. But the rut is just beginning and the contest does not last beyond this one brief encounter. In the next move the first bull turns and ambles down the meadow toward our hiding place. We are almost certain he has not seen us, but dare not wait. We move off at right angles to the bull's approach and almost immediately arrive at the edge of a flooded willow swamp. Without hesitation we slosh across, knee deep, expecting at every second the rush of a hairy brown monster from behind. But it does not come, and we emerge onto land again, dripping, and shivering in the aftermath of what might have been. When we judge ourselves past the meadow we angle back to the track, but soon hear bison grunting again and have to extend our detour. Eventually we strike another track which half a mile later rejoins the first.

Finally, near noon, we emerge from a grove with the full sweep of the Soapholes ahead. Nor does it stop at the park boundary half a mile away, but extends beyond like an estuary leading to the ocean. The Soapholes proper occupy the foreground; beyond, black terns hover and swoop over green waves of sedges. The pools of liquid mud, each with a central clump of bulrush, curve along the base of the steeper north edge of the basin. Peripherally the mud is crusted, but is everywhere punched deeply by the hooves of the animals who have come to lick the mineralized earth. Not all leave again. It must have been last autumn that the bull elk mired. Now his skeleton whitens in the sun.

The salts that saturate the clay and rime it with white percolate through gravel lenses to the surface. The most saline tolerant land plant is alkali grass, its grey-green tufts scattered over the trampled mud. In places, just before the meadow grasses begin, silverweed sprawls its bottle-green leaves that overlap again and again, mattressing the bare earth. A leaf overturned by the wind or a walking animal shows the white-felted underside, as pale as the elk skeleton. Where the silverweed thins, or is absent, patches of gumweed dominate. They are sturdy, enduring plants topped by coarse, deep yellow composite flowers, and the common name comes from the many, sticky recurved bracts that support the flower head like an embossed bowl.

We lunch just within the trees at the top of the slope, but nothing comes in the two hours we wait, drowsing away the hot summit of the day. Once a pair of goldfinches fly over; the terns continue quartering the marsh; alfalfa, pearl crescent, and fritillary butterflies point out the flowers and occasionally alight on the wet mud; dragonflies sift through the floating heat waves, a number in tandem mating pairs.

Finally we pull ourselves up and begin walking slowly back. The afternoon is cloudless and windless, hot almost to the point of wilting. Then the forest enfolds around us womb-like, but once we adapt it is to a different world, not a simpler, shielding one.

At the "bull sparring" meadow we are surprised to see the grass humped with brown forms, the whole herd lying placidly chewing their cud. We repeat our detour and on the far side of the swamp take up the main track once more. For perhaps a couple of miles it seems identical to the one we followed this morning. There are the clumps of jewelweed, the patches of agrimony and fireweed, seed capsules replacing blossoms on the lower half of the spikes, the drooping dogbanes, the scattered asters, ripe dewberries, raspberries, and currants. Red-eyed vireos sing above; a robin flies up, this time a juvenile; a downy woodpecker undulates away from the bole of an aspen. In a hollow filled with alder I catch the movement of a small bird, an alder flycatcher, true to its habitat, intent on its flycatching, so intent that its only sound is the mechanical clicking of its beak. I cannot find the wasp nest again, but am not concerned.

Suddenly, we step into a shrubby opening and immediately realize we are not where we thought we were. The track fades out. We carry on in the same direction, but are soon confronted by a wide impassible sedge fen with a small pond at its centre where half a dozen mallards are feeding. I recall that more ponds and at least one bog lie between us and the road. We return to the clearing and pick up a southwest trending bison trail; but it does not hold to its bearing long and we have to push through scraping, spider-webbed, head-high undergrowth. Once when I glance up my eyes fix on the grey lantern globe of a hornet nest several feet above. The black and white owners issue to and from the opening, and one descends to circle my head. I flinch and plunge ahead.

Without a compass we angle more west than south and eventually begin to skirt the edge of a very large bog. But now at least we have a familiar landmark. A moose trail follows the bog margin and although longer, is an easier route than a short-cut across the trackless ridges. At first the spruce are very dense, with oceans of darkness behind their waves of blue-green foliage. Then they thin out and Labrador tea foams up in the sunlit spaces. Under it are red-orange, scarlet, and hoary blue, the fruits of cloudberry, dwarf raspberry, and blueberry, like small sea animals in a mat of yellow-green seaweed. Summer swells to its annual girth of repletion.

August

Today is the second of August and for the first time since the solstice I am aware of the shortened daylight. The nights have drawn out by almost an hour and a half; the long ebb is gaining momentum.

But no intimation of the end of summer clouds this morning. The sun rears into a clear sky and is hot within an hour. Soon a light breeze flexes through the foliage and washes over my face and arms.

Everywhere I am met with evidence of fulfillment—seeds and fruits ripening, hooded mushrooms and brown moss capsules ready with spores, ponds and lakeshores clogged with water plants and surfaced with duckweed and algae, immature birds outnumbering adults, spike-horned bison calves, ants fluttering skyward in brief-winged mating flights, wasp and bee numbers swollen by generations of workers, wolf spiders under smothers of spiderlings. . . . . Molten sun, copious abundance, and voluptuous contours alchemize into a golden sensation of languid indolence. Time now as no other to wander without fixed destination, to absorb rather than probe and analyse.

Night-dampened cool lies in the shadow of spruce. A blue jay I never see calls repeatedly—first, squeaky phrases that sound like a rusty hand pump, then several rending *jeeee*'s. After these, silence. Chickadees move within my orbit of hearing and soon they are close feeding, confining their search almost entirely to twigs and side branches. Suddenly one hammers at a green twig. In a few seconds part of the twig seems to peel away and I realize the chickadee is tugging at a long, slender sphinx moth caterpillar. Exerting all its strength, the bird overcomes the insect's grip and flies off to a horizontal twig with the caterpillar dangling limply from its beak like a pendant dewlap. On its perch the chickadee holds down the caterpillar with both feet and pecks over the whole body for several seconds. Then it concentrates on the head end and is soon able to peck off small fragments from the now dead caterpillar. The chickadee flies to another perch, again carrying the insect in its beak; transfers it to a foothold, and pecks off more pieces. In less than a minute it changes perch again, and yet again. Its companions take no interest in its manoeuvers and I am puzzled by the chickadee's frequent moves. When I last see the bird and its prey the caterpillar has been half consumed.

On the ground under spruce pale-green cone scales glimmer. The seeds are not yet ripe, but red squirrels are already turning to this, their staple food. Whole cones, unopened and fallen prematurely, also lie on the ground, unclaimed. Squirrels seem to prefer to cut cones from the tree, perhaps because experience has taught them that these contain the most and largest seeds.

August belongs to the insects and spiders, and they seem to fill every niche to overflowing. Doubtless too, I am more aware of them because bird and mammal activity has declined. So many spider webs are strung across this narrow trail that at almost every step I destroy at least one. To attempt to avoid a visible web is to blunder into another unseen. The majority are suspended in shrubs at a height of three to four feet, but occasionally they are almost on the ground or glint high up in dead trees. Each web spans a square foot or more of space, yet most of the entangled insects are very small midges, muscid flies, and a few crane flies; only once do I see a dragonfly, its wings crumpled and the whole carcass mummified in silk. Rarely is a spider out on its web unless dealing with freshly trapped prey. More often I find the owners esconced in their curled leaf retreats near a corner of the web. The majority of these orb spiders are mature females with plump bodies that would almost cover my thumbnail, and long legs that double their size. Some are orange and sepia, others, more numerous, are a variegated greyish-buff with red-and-white or black-and-white banded legs. I begin to wish I had not sought out the webs' makers, for every so often I imagine one crawling down my shirt collar. That, however, is the last place they would choose to be and I always find the suggestive object to be a piece of leaf or twig.

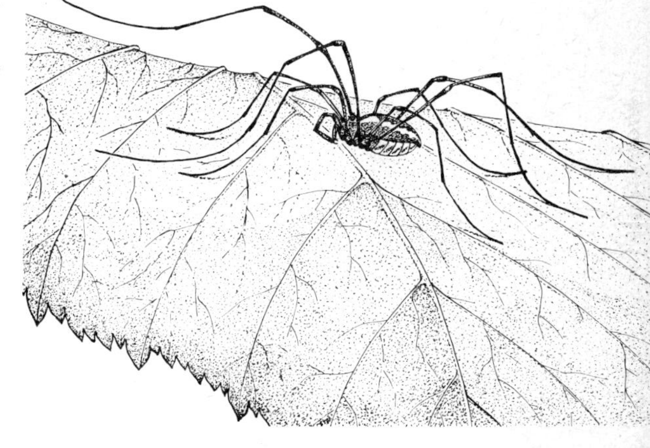

Daddy-long-legs walk splay-legged on the decks of leaves or bask in sun-spots. With their seed-bulbous bodies suspended on frail legs and their slow dreamy movements, they impress me as sidelines that should have passed into oblivion long ago. Yet they and their immediate ancestors have existed for at least four hundred million years and their present abundance augurs well for the future.

A bark chip on a rose leaflet moves—a jumping spider, grey with black stencilling, not more than a quarter-inch long. This family is well named, for their movements are always jerky, and they stalk and pounce on small insects without the assistance of a webbed snare. Suddenly this spider jumps, landing again on a grass blade a few inches below the rose leaf. It remains motionless for well over a minute before proceeding to the tip of the blade and finally disappearing on its underside. I see quite a few more jumping spiders during the day because I deliberately look for them. Lacking advertisement they are less conspicuous than any of the web-weaving species.

Pale buff blotches coin many hazel leaves, and within each circle, just visible, is a sawfly larva chewing away the leaf's substance. Before they mature and desert their mines to pupate large numbers will have been eaten by migrating songbirds, especially warblers, who slit open the thin covering and pluck out the soft grub.

More fruits are ripening to glowing reds—crimson-dotted capsules on mayflowers and Solomon's seal, handfuls of vermilion on the bush cranberries, the familiar puce of raspberries, and the slender maroon pods of dogbanes and fireweeds.

A pond shines through the trees ahead. As I near it an adult red-tailed hawk falls from the top of a dead tree and flies along the shore screaming. Its moult has begun, and missing tail and wing feathers give it an unkempt appearance, although its flight is normal. Instead of leaving the pond it moves from perch to perch with *skeeeeee*'s of alarm as I walk towards it, and eventually circles back to its starting point. Perhaps it has a nest nearby, or just-fledged young. Looking over the water into the sun, I identify with certainty only two adult red-necked grebes; the few ducks look like widgeons, mallards, and teal. A song sparrow trickles out his refrain, once only.

Long before mid-day the columned forest is like an empty chapel, calm and reflective. Most birds are beginning their pre-migratory moult and during it are secretive and quiet. In half an hour I see only a juvenile white-breasted nuthatch methodically exploring tree trunks, both live and dead, and two hairy wood-peckers likewise seeking food; hear only a scuttling in the leaves that could be a bird, a mouse, or a chipmunk.

Upland openings are now the nodes of activity. New flowers replace those spent or nearly so, spreading in garden-like profusion as if they know they are the last and their time to set and mature seed is short. Buckbrush continues to open flowers even as the fruits of the closely related snowberry are whitening to ripeness. Robins, cedar waxwings, three red-eyed vireos, and a catbird are feasting on the saskatoons, pin cherries, and raspberries. Choke cherries will not be edible until their matte red transforms to obsidian black, and rose hips not until they are totally red. Most of the acrid-pungent fruits of buffaloberry are gone and next year's flower clusters have appeared.

Some grasshoppers shrill almost incessantly; others arc away from my feet in castanet flight. Dragonflies and damselflies shimmer back and forth through the open air. Butterflies flicker from blossom to blossom, showing distinct preference for asters, goldenrods, and thistles. An occasional white admiral is faded and tattered, probably a lingering survivor from early spring who will be claimed by the first frost. Bumblebees, white-faced hornets, and hover flies compete with butterflies and each other for space at the most favoured flowers.

I cross a disused track. In the soft, rain-flattened earth are the cleanly incised prints of a white-tailed deer and her fawn together with the mingled, overlapping ones of half a dozen elk, cows and calves. Much of the deers' sylph-like character is expressed in their small, tapering hooves and mincing steps, just as the elks' broad, blunt prints and longer stride reflect their heavy bulk. Here also I find a coyote scat, composed entirely of saskatoon seeds and skins.

At the kestrel slope I think all the falcons have left until the female flies over. Gaps in her tail and wings show that she too is moulting, for the losses are symmetrical. But like most birds she will never become flightless, nor even severely hampered. The kestrel skates on with alternating series of rapid shallow wingbeats and short level glides which soon take her out of sight. The nest tree has served its purpose and will have no special attraction again until next spring.

But I see a movement near its base. I finally discern a pileated woodpecker in the shadows, half hidden by shrub foliage. However, she is only using the tree for support as she stretches her neck to pick pin cherries one by one. In late summer, autumn and winter fruits and large seeds are often eaten by several species of woodpeckers.

Shrubs near the top of the slope where the forest begins shake with movement. A rose-breasted grosbeak planes down into a saskatoon, joining other gourmands. Half a dozen cedar waxwings, all adults, occupy the upper half of one pin cherry. Gracefully, sibilant with pleasure, they pluck the glossy red fruits. Most of the other participants are silent: several robins, adults and juveniles, indifferent to security, choose the most exposed and ripest berries; a blue jay, feathers sleeked demurely, slides from branch to branch; a female oriole, inconspicuous in wood-brown and muted yellow-orange, picks here and there; a family of red-eyed vireos, with a single warbling vireo near them, fuss about the twigs; three catbirds and a veery select the lowermost berries under a shield of foliage; and two white-throated sparrows, perhaps more, pick up fallen fruits on the ground below. Occasionally two birds, of the same or different species, reach for one berry simultaneously, but the scuffle of raised wings and feathers, gaping beak and sharp note of protest is momentary. The bounty is spread over a large area and the birds rarely come close enough to quarrel.

A chipmunk has one saskatoon to itself. Unlike the birds it does not swallow whole berries as it picks them, but cuts each one free with its incisors, then carries the food to the ground where, sitting, it holds it in the front paws. The chipmunk seems interested only in the seeds, and once when it bounds off with cheeks bulging I inspect its dining site, finding it littered with discarded skins still containing some pulp.

On the way home I examine a lagoon almost cut off from the rest of the lake by a raised bar of sedges, cattails, and young willows. I note the usual leeches, water insects, snails, and fingerling sticklebacks, but what really attracts my attention is a flocculent, semi-opaque substance encrusting several inches of a waterlogged stick. Such is the appearance of a mature, fresh-water sponge, an almost structureless, greenish-dun mass which over the summer has become further camouflaged by leaf and twig fragments, as well as the occasional dead insect. At the edge of the water coltsfoot and marsh marigold leaves are tropically lush. On the bar the cattail lances are spike-tipped now that the stamens have withered away. The browning seed cylinders are suddenly conspicuous.

I turn onto the road, keeping back from the shore where ducks are resting on partly submerged logs. Some distance ahead a dark animal thrusts out of the concealing sedges and angles across the road toward me. It is not like anything I have seen before, low to the ground and with a flopping, sinuous gait. Then it turns slightly, almost face on, and I realize the creature is really two—a mink dragging a full-grown mallard or widgeon as big as itself. Yet so swift and silent was the kill that the other ducks along the shore still sleep, or preen unconcernedly.

This day is one of domineering sky, the piled, billowing, densely textured clouds vastly more important and substantial than the land they subjugate. So high-turreted are they that blue rarely shows between, yet intermittently the sun pierces a gap, only to be extinguished a few seconds later. The land and water are brilliant then sombre, warm then cool, expanding then contracting. The breeze too is intermittent and vascillating.

By mid-afternoon the entire span from horizon to horizon is freighted with dark-bellied, pregnant herds, as I imagine bison once obscured the prairie grass. The breeze has stilled, and the clouds roll on in silence.

A blur passes before an island in the lake, becomes denser as it slowly nears. Now my ears are filled with a sound like millions of angry bees, a hissing, seething, pricking sensation that intensifies rapidly. But for a while I still do not connect the sounds with the spread of dimpled water. Then I see it too is inching closer. The curtain of gauze sweeps in, countless tiny white pellets pummelling the water, the friction of their impact generating the strange, insect-like humming. The hail drives down on the ground, bouncing, rustles off leaves, gathers on my jacket, pin-stings my skin. I retreat under a shaggy balsam poplar. In a few minutes slushy rain mingles with the hail and soon completely replaces it. Raindrops plop into the lake, spatter onto high leaves and drip to lower.

The rain lingers for half an hour; then cloud tops pull apart and dazzle white against ultramarine. However, the clearing is temporary. Soon the clouds writhe close again, and embrace. The damp air cools quickly and the breeze off the lake is raw. Before night folds in the landscape more showers trail over, each separated by an interval long enough that I look expectantly for signs of sunset.

❧ ❧ ❧ ❧ ❧ ❧ ❧

The recent showers have had a rejuvenating effect. Aging leaves are a brighter green in the early light, glinting with dew. But no amount of rain can wash away the insect blemishes, although the loss of botanical perfection has been the gain of zoological interest. Plant foliage is an integral part of the whole ecosystem, existing not only to nourish the plant body, to shade and to decorate, but also to feed and shelter unnumbered insects, as well as moulds, bacteria, and viruses. The cumulative effect is that now few leaves are pristine. A single leaf, if it is large, may have one or more galls, an internal mine, notches cut out by a caterpillar, and a patch of white mildew. Whole stands of cow parsnip have their basal leaves whitened by fly mines, as many as two dozen larvae in a leaf. Occasionally I come upon an aspen whose foliage from a distance looks grey, the result of moth caterpillars which, in winding around and around the leaves' interior eat most of the green cells. Then there are the leaf rollers, many the larvae of tortricid moths, which do their work with admirable neatness. One day in late July I watched a chipmunk in a tangle of red-osier methodically testing each rolled up leaf with a single bite. It did no more to empty ones, but from the others extracted and ate the larva or pupa.

I pass a hazel where gobs of white foam are gathered at the base of several leaves. Looking carefully on their underside I find spittlebugs undergoing their final moult. One adult, fully expanded and a uniform leaf green, rests quietly on its discarded skin waiting for its new tissues to harden. Others, a weak yellow-green, are beginning the struggle out of their split nymphal cases. Only one fat nymph is still producing spittle—the bubbles emerge from the tip of its abdomen as it swings slowly back and forth.

A male oriole flashes through the tree crowns, whistles once melodiously, and is gone. Then a red-eyed vireo sings, infrequently but persistently. Each time I think I have heard its last utterance the querulous phrases come down once again. Robins are moving restlessly through the forest. I come upon them singly, more immatures than adults. They make short flights, perch, call softly a few times and fly on.

Five grackles fly over the lake from a dead poplar. A song sparrow drops down from a willow. A LeConte's sparrow sings once from a sedge-choked bay. In a protected cove the water is sheeted with moulted duck feathers as if a pillow had been shaken out. Far out are the ducks themselves, broadcast over the water and here and there gathered into loose rafts that blot the skylit lake like ink smudges.

I scan the little indentation ahead, partially cut off by screening bulrushes. The water is still, dark and patterned with the reflections of sedges, shrubs, tree trunks, branches, and foliage. The red-necked grebes who nested here have left and their platform no longer looks like a deliberately fashioned object. The place seems empty of birds. Then I notice an unusual projection on a fallen tree reclining into the water. Through binoculars it becomes an immature black-crowned night heron, finely streaked and spotted in neutral umber and white, knobby legs pale greenish-ochre, stiletto beak dark. In many ways, including body proportions, it is very like a bittern, but lacks the warm buffy overtones and black sideburns framing the white throat. The heron moves slowly and lets itself down into the shallow water. There it stands against the sedges absolutely still. When I look for it unaided, I cannot find it; it is one with its background. Finally the young heron begins stalking with great care, lifting each foot clear of the water and putting it down again with barely a ripple, body leaning forward in anticipation, beak angled down. Several times it stabs into the water, but only once does it bring up a wriggling, silvery stickleback. Continuing its fishing, the heron in time moves out of sight. I walk on. Almost at once an adult black-crowned night heron flies heavily from a shoreline tree with a hoarse squawk. It must have been there all the time, perhaps supervising its offspring's fishing. Beating its wings rapidly the young bird catches up to the adult, then both fly to a nearby island and disappear around the far side.

The bog and its flanking white spruce draw me into their sanctum calm. A sapsucker taps without pause, reaming out a little well in the trunk of a white birch. In the gangling-tall alder a female and an immature myrtle warbler feed leisurely, flitting from twig to twig, peering at and under leaves, now and then uttering conversational *chup*'s. In the same alders, but close to and on the damp ground, five juncoes, two adults and three streaked juveniles, hop about and scratch in the matted leaves. A young bird hops close to a parent and wheezes in supplication but, ignored, it stops in a few seconds and returns to its own searching.

Half a dozen black-capped chickadees lighten the recesses of the bog with their calls and slowly approach the moat. From behind three boreal chickadees come up through the white spruce lisping huskily. The two groups move closer until separated by little more than thirty feet of moat. They call more intensely and frequently, and soon the black-caps fade into the poplar forest beyond the bog. For the moment the boreal chickadees remain feeding in the white spruce. The encounter seems to have been mild enough, yet I wonder to what extent the two species are competitive. From all my observations here spruce stands are the boreal's exclusive habitat. On the other hand black-caps frequent poplar, mixedwood, and pure spruce, seeming to prefer the first for nesting, and the last as a refuge in bitter winter weather, but at other times spreading through all three almost equally.

Underfoot the moat is firm, yet springy with the inherent resilience of peat. At the edge of the black spruce many ephemeral mushrooms rise through the mosses, their caps delicately shaded in tints ranging from dark chestnut to umber, ochre, pale buff, ivory, even stark white. A large wood frog hops away from under the heavy, leathery brown dome of a bolete. A few steps farther on I disturb a toad almost as vividly green as the mosses. Clumps of spiky, stiff clubmoss are topped by smooth, pale green strobili from which dust-fine spores will sift in the autumn. A crane fly drifts past with its paddling flight and dangling legs, caught in a streak of accenting sunlight. Also heightened by the sun are the many sheetweb spiders' silken traps which bridge narrow gaps between twigs. A densely woven, yet not opaque sheet forms a flat roof on the underside of which the small spider rests. Subtended from this a three-dimensional network of intertwined strands snares winged aphids, midges and other tiny flies.

The interior of the bog gleams brightly sun-spangled, but most of the plants there also grow at the outer edge where I stand. Now that the Labrador tea has been thinned by the loss of last year's leaves, turned orange and tawny weeks ahead of autumn, herbs and trailing shrubs assume more importance—the reddening capsules of three-leaved Solomon's seal, orange cloudberries, scarlet dwarf raspberries, crimson bog and dwarf cranberries. Yellowing cloudberry leaves add splashes of pigment sun, while fresh ones just sprouting will ensure growth until deep, killing frosts arrive. In the moat marsh marigolds are also putting out new leaves.

As I leave the bog, bending under the sagging limbs of a white spruce, I nearly trample a patch of fairy-sized mushrooms. I drop to my knees and put my head to the needle-strawed ground. Flat circular tan caps little more than a half-inch in diameter sit atop two-inch black wire stems. Some of the caps are wizened. Although they look like typical mushrooms when fresh, most kinds of Marasmius endure for months after they have shed their spores. The stem and cap dry without decaying, and after being wetted by rain expand to their former size.

A balsam poplar leaf spirals to the ground. When I pick it up I see that a large, globular gall on its mid-vein has been freshly sliced open. More recently fallen leaves lie on the trail; all with galls half and more eaten. In some I find a few of the galls' makers—small green aphids. An animal scratches and rustles in the tree above. I soon locate a red squirrel moving from twig to twig, branch to branch nipping off galls one by one, and occasionally hanging by its hind feet to reach a cluster of leaves. The galls and their contents must be delicious, for the squirrel takes no notice of me and is still feeding avidly when I leave two or three minutes later.

I emerge into the brightness of an old clearing now a welter of raspberries, roses, currants, cherries, saskatoons, and low-bush cranberries. Simultaneously I am barraged by fear-mingled yelps of rage. The source is almost certainly a hawk, but I cannot find it. I step forward. A small bird flies and perches in the next aspen where I see it clearly—dark grey-brown above, chestnut bars on white below—a sharp-shinned hawk. It bends forward on the branch, lean, like a greyhound at the starting gate. Scarcely larger than a blue jay, it is as predatory as an eagle, from curving raptorial beak to tensed grappling claws. It cries again, this time firing a long train of cackles in my direction. I move and suddenly the hawk flies at me, swooping over my head, curving up to land in a birch just behind. It could have young on the point of fledging, but the nest is too well concealed to be found readily. The hawk stoops at me twice more before I am beyond the circle of its defence, and continues calling for some time after.

I start across a grass meadow under a fleet of full-sailed cumulus, on the way to a tiny pothole pond in the forest beyond. The meadow is empty, but brims with a comfortable quiet. Dragonflies and a few damselflies patrol the air; bumblebees purr from blossom to blossom; the occasional fritillary and alfalfa butterfly flickers up from the dandelions edging bison wallows, or thistles and goldenrods in the turf. I am halfway across when bison grunting belches from the trees ahead. I stop, my heart pounding in my throat, a wave of fear curling through my body. I am completely vulnerable, the nearest trees as far away as those from which the grunting issues. I strain to see into the shadows—is my adversary a bull? or just an excited cow? or perhaps not an adversary at all, the grunting directed at another bison? but I can make out nothing behind the sun-washed saplings and shrubs at the edge of the forest. The grunting continues and seems to develop a gurgling undertone. I back up slowly, hoping my actions will imply that I am yielding to the challenge without contest. To continue standing or to sprint to cover might invite a charge—aroused animals have hair-trigger reactions that one wrong move can precipitate into violence. I continue backing for a long way, then turn and saunter, I hope with an air of indifference, to the trees. Once among them I look back. The meadow remains empty, and the grunting has stopped.

I sit out on the point in the early evening absorbing the declining sun's fireside glow. A dragonfly alights on my knee. It is small and slender, clear-winged, and with a light claret body. The abdominal segments pump like little bellows as it inhales and exhales through its spiracles. The dragonfly appears to be resting, but while this may be true to a degree, it is at the same time fully alert, frequently swivelling its head up and down or tilting it sideways to cover the full orbit of sky, land, and water. Suddenly it flies and in moments I lose it among others.

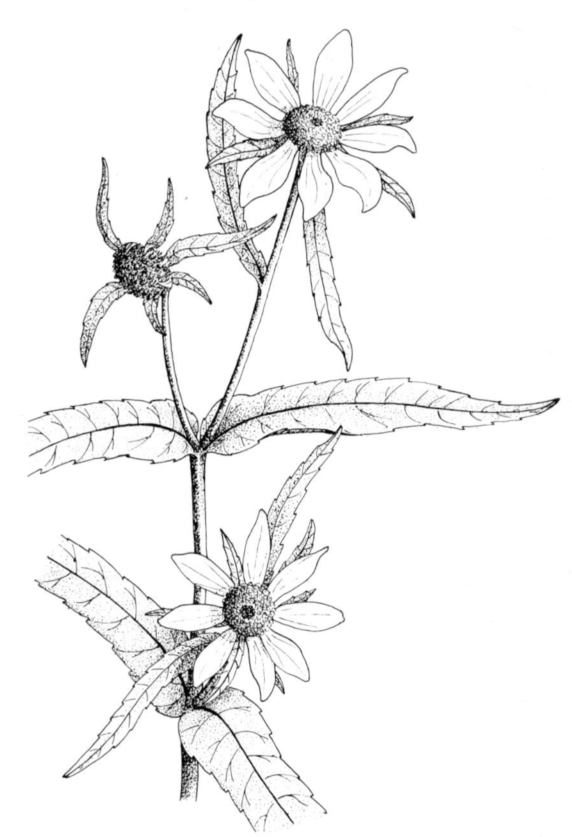

As the light dims the arrowhead flowers on the east side of the point brighten like little moons. Beside them the pink columns of water smartweed turn smoky. The first beggarticks have opened. Both ray and disc flowers are intense yellow, but the ray flowers are broad and few in number. They are the last plants to begin blooming.

The dwindling clouds colour flaming rose as I walk back to the house. On the garage roof three phoebes are silhouetted. Although long finished with their eaves' nest, they have shown no inclination to leave the site. Coyotes ululate briefly, and after they cease a distant white-throated sparrow flutes thinly.

The cycle began five days ago when a phalanx of violent thunder and hail storms thrust eastwards across the entire province. When the cloud ranks broke behind the front a day later the drenched land steamed in the sun. Since then each day has begun with a pristine sky, but by mid-morning young cumulus flock like lambs in a blue pasture. Through the rest of the day they grow and thicken, maturing into heavy-fleeced rams by late afternoon. The sky is crowded with them, but they do not merge. At times I stand bathed in sunlight watching one a few miles away parading downwind and trailing the smoke of drilling rain. Sunset is lurid, the cloud-rams purple with rage. They have reached the zenith of their stature and power. Now lightning flickers, flares, and thunder rumbles as the first charges are run. Sometimes a particular storm comes no closer. But inevitably, usually before midnight, one monster rolls overhead. The night is shattered, blasted like a battlefield under seige. Bullet hail drums on the house, lashes the trees; rain is flung like shot against the windows. Branches are ripped off trees. In an hour or so the storm passes and night finds its rest again. No new storms will arise until tomorrow's sun recycles the rain, each night-deluge feeding the next day-clouds. Yet not all of the rain evaporates again and gradually the storms are diminishing in intensity.

This night, the sixth, is impenetrably dark, wet with an early dew, whole constellations of stars blanked out by cloud, but its peace inviolate.

For the first time since the end of May the night temperature has dropped below 40 degrees—this morning it reads 37. Crisp. A warning on the eighteenth of August.

The sun rises quickly into an empty sky and soon dispells the chill that is like a bad dream. A breeze ripples the lake under the grebes, ducks and gulls; stirs the leaves; disperses the smudges of fog along the lake and in hollows.

I pass a bog on the way to a beaver pond. The seed cones of white spruce, black spruce, and larch, now all a uniform tan, stand out distinctly in the green shrouding of the trees. White birch and alder cones hang green and incomplete-looking, like premature foetuses. A red squirrel detects my passing and chitters a "no trespass."

I stop. A cow moose and her calf block the trail. Both turn their angular, roman-nosed heads, and cup their oversized ears toward me. I have expected to meet them sometime as they have been in this area most of the summer and have become accustomed to people. Even so, is a hundred feet too close? The cow's ears are still forward, but I back up slowly and look at the pair obliquely, turning my head as though interested in something at the side of the trail. The seconds drag, then the moose amble off. As I wait for them to move well away an immature yellow warbler shifts through the shrubs intent on finding food, ignoring me. I watch it slit open several sawfly blisters on the hazel, and later pick a small green caterpillar from a green leaf petiole. I have the impression that most often the caterpillars I see in the beaks of birds are slender, smooth, and green. This would seem to negate the value of camouflage, but of course, many more might be found and eaten if they contrasted with leaves; and it is true that few birds eat large, brightly-hued or excessively hairy caterpillars whose conspicuousness advertises their unpalatability.

I edge forward and cross the dry end of a swale. The moose are now at least a hundred yards away, half-submerged in sedges and browsing the bordering willows. They must hear or see me for they turn their heads, but are only momentarily distracted.

At the beaver pond I find a comfortable log from which I have a broad view. I disturb a song sparrow and it flits away through the shrubs flicking its tail and *chink*-ing; however, after a few minutes' stillness I become part of the scene.

Something is missing—the bustle of red-winged blackbirds. I have not noticed the gradual decline in their numbers which must have been taking place over many days, but now that they have gone and I am aware, the cattails and willows seem forlornly vacant. Once the young are fully fledged the red-wings gather in large flocks which feed on open uplands where the birds walk over the ground picking up insects and seeds.

Four tree swallows approach from the far side of the pond, looping, circling, skimming the water in smooth glides like ice skaters. In a minute or two they leave.

I hear familiar trilling and nine small birds sweep into the top of a dead tree. To my surprise they are Bohemian rather than cedar waxwings. All are adults with the same debonair appearance as their slightly smaller relatives, but are greyer and have rusty instead of white under tail coverts. The Bohemian waxwing is a boreal forest and mountain bird in the summer, and while a regular winter visitor, rarely appears in this area before November. These may be birds set loose early by nesting failure. Their wandering is not over however; they soon take wing and I never see them again.

Stretched between the points of its uplifted wings and long-reaching toes, a solitary sandpiper alights on a stump. Being immature it lacks the white spotting of the adult and has greenish legs instead of black. For nearly five minutes it preens with complete absorption, its long needle beak penetrating the feathers deftly, cleaning and rearranging. Then, with undulating, snake-like oscillations of its head, neck and beak it smooths oil over every surface. A short flight and it drops into shallow water near the dam. Now for some time it wades, often up to its belly, dipping here, dipping there as it catches insects. It makes no sound.

The song sparrow comes back, quiet now, feathers smoothed. It hops about the shrubs and once picks a small insect from a twig. Then it moves to the ground and I can hear it rustling in the dead leaves. After that I lose track of its whereabouts.

Ducks are lined up on logs rising out of the water. From a distance their cryptic colours give them the appearance of burls, part of the trees themselves. Only those feeding in the water look like real ducks. I examine each one individually, through binoculars, and after long, intent study separate species by the size, shape, and colour of the head and beak, together with minor, yet consistent differences in pattern and tone of the body plumage. However, the distinctions between adult hens and full-grown immatures are too subtle for me.

Nasal, petulant croaking see-saws through the trees and within seconds two crows appear close together, flapping heavily. They land on a branch and now it is obvious that only one is making all the noise. The other holds something with its feet, probably the remains of an animal, and tearing off pieces, offers them to what must be a young crow, although both seem identical.

The sandpiper has gone and I walk along the overflow channel below the dam. The heat has revived the dragonflies whose quickened wings twitch them on sunlit sorties over the water, over the sedges, and a few, into the sundappled forest. Ahead an alder flycatcher repeats its abupt *fe-bee* over and over, while closer, a yellowthroat warbler fusses in the willows warning all others with its *tchik*, *tchik*. Soon it is joined by a Connecticut warbler uttering loud *chip*'s that seem to have been expelled by force. Yet it continues its feeding, moving from twig to twig and picking insects from the leaves as well as snapping at ones in flight.

A gap in the shrubs allows me to glance over the breadth of the sedge fen. On the far side a bull moose suddenly rises out of the green. He pauses, high head crowned with broad, handlebar antlers still in velvet; then swings his head away, strides off and is enveloped by forest.

On my way back to the pond a little knot of pine siskins passes over muttering softly. A white-throated sparrow flushes, pauses on a twig, flies on. A ruffed grouse whirrs away.

Under the high noon sun the ruffled water glitters like shards of glass. I find the glare of the pond repellent and move to sit in the restful green radiance of the undergrowth. A toad, tailless but not more than an inch long, hops away from my crushing boots. What does it think of the monster looming above? or is it aware only of my feet? They alone are gross enough. However, after a few hops the toad stops under a drooping leaf that nearly touches the ground.

An old grass seed head has its spikelets pulled together, curved under tightly like a fern fiddlehead and held in place with silk. A tiny hole makes a dark circle on the inner end. I tap the dry stem lightly. Immediately a grey and black jumping spider scurries out and drops two feet to the ground. Although these spiders spin no web, like all species they can produce enough silk to fashion a secure retreat.

During the past few days butterflies have suddenly become less numerous. Yet, crossing a meadow in the afternoon I see several fritillaries, checkered whites and white admirals feeding at flowers, and put up a mourning cloak and an alfalfa from a drying puddle. Bumblebees, hornets and hover flies still clutter flowers as if a lifetime of living lay ahead.

I woke to rain during the night, a gentle dripping without wind. The overcast lingers into the day, but the stratified clouds, banded light and dark like the paper of a wasp nest, have lifted. A breeze shivers in from the northwest.

The abrupt change in weather has brought down the first wave of migrating birds. A hundred garrulous coot, adults and immatures, dot the bay behind the house. They dive repeatedly and bring up long streamers of pondweed which they peck and shake vigorously, snipping off leaves and seedheads, and incidentally swallowing snails and insects. As soon as a bird surfaces its nearest neighbours converge on it and help to demolish the pondweed, usually without objection from the owner.

The point, which has been almost deserted this month, is now alive with birds. Several orange-crowned warblers mix with yellow-cast immature Tennessees and magnolias. Of the last, one is a bright yellow and black male. A few least flycatchers move through the saplings and trees, for the most part somewhat higher than the warblers. They concentrate on flying insects, beak-clicking interrupted now and then by sharp, assertive *kwut*'s. Occasional *tseep* notes maintain contact between the warblers as they work through the undergrowth examining leaves, extracting sawfly larvae, ascending, descending, constantly in motion. I look through the group again. A female redstart fans her tail with a sudden spread of yellow; flies, and reveals a similar yellow wing-swath. A single ruby-crowned kinglet appears to be travelling with the warblers. Smaller than they, it is also plumper, and has a shorter beak. A narrow white ring accentuates the shoe-button eye, excising it from the muted olive-brown plumage that pales to ivory on the underparts. After the cool night all the birds feed eagerly. They are following the lakeshore and I soon leave them behind as I walk.

It is afternoon before I encounter another warbler flock. Many birds arriving in spring first appear over a broad front. One day there are none, the next they are everywhere. Autumn migration is different. After the dissolution of nesting territories birds' movements tend to be irregular. Some, like the blackbirds and swallows, soon gather into large, wandering flocks. Others collect in small groups, often of several closely related species, and begin to meander south. They stop to rest and feed when the need or weather drives them, and thus may overfly some places where I expect to find them, or have done so in other years.

These afternoon warblers are in willows and alders ringing a bog. A sleek, clove-brown waterthrush walks on the damp leafy ground, in and out of sight among coltsfoot and marsh marigolds. It teeters a little, like a spotted sandpiper, and several times I see it turn over dead leaves. In the shrubs above I count four Wilson's warblers—two females, one adult, and one immature male; two redstarts—both immatures with pink-flushed throat and upper breast in addition to the familiar markings; and three mourning warblers—probably immatures since their grey hoods are tinged with olive. Each bird is preoccupied with feeding. Intermittent clicking leads me to a wood pewee at the very top of an alder. The clouds have passed and released the sun. In the warmth many more insects are flying.

The warblers' gentle, forgetful calls are suddenly buried under an avalanche of jumbled stutters, twitters, and husky *dee-dee*'s that rolls in advance of a band of several chickadees. They move in from the poplar forest, through shrubs, and mingle briefly with the warblers; then continue on across the bog. The warblers seem little affected by the rush of chickadees, but now I cannot find the pewee. A bird hops on the ground—a glimpse of its brown-buff striped head and white throat is enough to identify a white-throated sparrow. A longer view a few moments later shows the light breast streaking of the immature.

I walk home through poplar forest. Several robins weave in flight between the columnar trunks, above the shrubs and below the crowns. Their high, drawn-out calls drag in the air behind. A flicker in the top of a dead, blackened tree rings out several *keee-uw*'s, the intervals of silence long, as though it were waiting for a reply. I listen too, but no response comes. A Swainson's thrush flushes from the trail ahead, its earth-brown back invisible until the moment of flight.

Under the lilac-veiled asters, mayflower leaves have turned to pale papyrus and over half the crimson berries are gone. Tan and yellow leaved sarsaparillas, foreign in the general green, are finished with summer long before the first frost. Most pods of fairy bells are yellow like the discoloured patches on the aging leaves; others are tinting scarlet, close to ripeness. The tops of rose hips glow with the red of sunburned skin, but are still green underneath. Baneberries have been flawed by grouse, deermice, and perhaps others who, unlike us, eat the fruits with impunity. At the edge of the trail northern bedstraw is pyramided with a pale echo of its floral white. Stiff white hairs fur each of the many round, magenta seed capsules. In the sunlight they shine as a myriad of haloes.

A blue jay glides down into a hazel like a shaft of sky, then disappears among the leaves. Nearer, a second jay does the same, but

remains in sight. It cocks its head from side to side and moves from twig to twig until it finds a cluster of nuts, mature but green. Grasping the snouted husk near the base with its beak, the jay pulls the whole thing free with a few quick jerks, flies to a large horizontal branch, and pins down each end of the husk with its feet. After several sharp pecks on the bulbous nut the jay is able to pull off a strip of the husk. Further jabbing enlarges the opening over the kernel; then the pale sphere is poised in the bird's open beak momentarily before being tipped back and swallowed, not without effort. The jay picks and opens the next nut in the same way as the first. However, now it holds down the kernel and splits it in half, then breaks off fragments and gulps them quickly. But this more complicated treatment ends with half of the kernel falling to the ground. The jay does not attempt to retrieve it, but flies down and plucks a new nut. This time it reverts to swallowing the nutmeat whole. The other jay reappears and, when I watch it for a while, I see that it treats the hazelnuts exactly as does its companion. Yet this bird may be more practised since it manages to break up and eat more than one kernel without dropping any.

Along meadow edges, road margins, and on shrub slopes the first goldenrods are going to seed, the last fireweed blossoms can be counted, and choke cherries darken as their leaves begin to colour orange, scarlet and yellow.

September

Last night went clear to infinity. Stars webbed the void, a silver net cast over eternity. This morning, minutes after sunrise, dew prisms the light, but I scrape away a rime of frost from leaves in shadow. However, the forest has kept the cold at bay, and the undergrowth leaves are dew-wet.

By the time I am out again after breakfast, carrying a sandwich lunch, the sun is almost summer on this the first day of September. A song sparrow *chink*'s in the willows by the water. The bay behind the house is thick with feeding ducks and coots. Beyond them ring-billed and Franklin's gulls float silently, preening. Along all the shores, near and far, many balsam poplar have bronzed, a tone that seems to lie below the leaf surface, shadowed by green. Occasional yellow candleflames mark the first changing aspen.

I take the trail along the north shore of the lake. For some distance it provides only glimmers of the water, islands, bulrush beds, and sky as it curves up over knolls of mature poplar forest, skirts wet depressions and bays, and bridges trickles.

A red-eyed vireo sings several phrases, then silences and moves on without my seeing it.

During the past week most rose hips have ripened; fairy bell pods glow today like scarlet velvet purses; bush-cranberries are glossy vermilion; unfrocked, the mace-headed water arums are spiked with rubies; red-osier dogwoods and snowberries present clusters of virginal white; the last of the raspberries are soft, plum-red, juicy. Close to the ground bunchberries burn in the dimness like stone-age fires in cave mouths. Solomon's seal and oak fern, like the mayflower, have thinned to frail parchment before withering, the walking dead. Signs of retrenchment are everywhere. First leaves are now the oldest; have lived their allotted span—are yellowing on hazels, willows, birch; blazing red, orange, and yellow on old rose canes; glowing crimson on the bush-cranberries. Amid this senescence, new spring-green leaves on the wintergreens, bishop's cap, and twinflower are an affirmation of continuity.

At the first seep jewelweeds still flower, with buds to come as well as ruptured pods. A bumblebee drones in and lands on a blossom. However, instead of pushing open the mouth it immediately crawls to the pouch at the rear and bites into the base of the recurved spur, for it is too large to enter by the mouth. Draining the first, the bee moves on to another flower and another, treating each similarly. It does not leave until it has drunk from five.

Teck . . . teck . . . teck . teck . teck . . teck, an adult male sapsucker pecks methodically on the trunk of a dead tree. A minute later he flies away. Soon an immature sapsucker arrives and pecks in approximately the same place. Then it too flies off, in the direction of the adult. Since sapsucker families remain together until migration, these two birds are probably related, and the tapping communicative.

The trail veers toward the lake and the long bridge across the filled-in bay. Two chickadees come close through the trees, flitting from twig to twig until one is less than five feet from my face. As usual they seem to be seeking food, but curiosity must be an element in their approach, for they soon recede back into the forest. I step onto the bridge and a LeConte's sparrow shoots up from the sedges, then drops down only twenty or thirty feet away. I see no more of it. In the middle of the bay cattails rise sheer above the sedges, green ribbon leaves yielding to the breeze, brown-headed lances straight and steady.

Among the mallards strewn over the water from shore to a narrow, hooked island not far away I find blue-winged teal, lesser scaups, goldeneyes, coots, and several eared grebes, either immatures or adults already in their soft, grey and white winter plumage. A kingfisher rattles from the island and I finally locate it in a dead tree at the water's edge. Even at a distance, and with the sun behind turning it to a silhouette, its heavy-headed, spear-beaked form is unmistakable. It rattles again, the call high-pitched and petulent.

Without apparent reason a chorus frog begins calling from the heart of the cattails. The creaking comes irregularly, at long intervals, and is prolonged like a record slowed to half its normal speed. The frog seems lost in reverie, anticipating the coming spring.

At the end of the bridge a song sparrow hops along a log, vanishes into a tangle of sedge, reappears on another log, and hops down it to the water. There it drinks—dipping, raising its head high, dipping, raising; altogether half a dozen times. Then it flies across the bay, low over the water, hurriedly. The open has many dangers. As I walk on, a house wren chatters briefly.

Spider webs are in disrepair, torn, shapeless shreds of white. Many spiders over-winter as eggs secured in intricately structured silk cocoons or as young spiders, and once the female enters the last stage of her life in late summer, egg laying, she usually stops eating. The abandoned webs soon disintegrate.

I lunch sitting on a high bank overlooking the west half of the lake. Ring-billed gulls, flying and swimming, are distant flecks of white. As by the bridge, mallards are the most abundant ducks, with a scattering of widgeons, blue-winged teal, scaups, golden-eyes, coots, and eared grebes. But now I add more species. Six adult Bonaparte's gulls float, preen, and peck at the water. Whited out, all that remains of their spring-black hoods is a dusky basal line. Soon that too will go and they will spend the winter with sooty "eyebrows" and "ear" patches. Three buffleheads swim among the Bonaparte's, fractionally smaller than the gulls. Two widely separated pairs of adult red-necked grebes swim and dive. All are beginning to show winter grey, although their necks are still predominantly rufous. A fifth, a striped-face juvenile slightly smaller than an adult, swims alone. A drake white-winged scoter speeds past almost skimming the water, wing-flashes signalling. Although scoters have a mid-summer moult like most ducks, the alternative plumage is almost identical to the nuptial one, both being black. A green-winged teal stands resting on a log by the shore. Suddenly it stretches its neck and turns to look behind. A mink's dark head of doom emerges from the green sedges. But the duck has seen in time, jumps off the log and skitters desperately over the water with feet and wings not yet ready for flight. The mink pulls back; the teal subsides on the water, but continues paddling rapidly away from land. A ring-necked duck swims into view trailed by six ducklings who cannot be over two weeks old. It seems incredibly late for such a young brood, yet they should be able to fly five and a half weeks from now, in the middle of October, with almost a month to spare before the lake freezes over completely. *Peet, peet . . . . . peet, peet*, a spotted sandpiper parallels the shore, wings alternately blurred and set rigid, yet its flight path curiously even. I search in vain for terns. Both species must have left, but at the moment I cannot recall precisely the day I last saw either.

A mourning cloak flutters down where the ground has been scuffed bare by human feet and stays there sunning, fanning its wings slowly. The bright light reveals the full wealth of its primary colours: red as a deep undertone maroon over most of the wings, yellow as a rich cream in the marginal bands, and blue as horizon sky in the single row of spots.

As I turn to go inland five cedar waxwings fly into a dead tree, trilling gently. Four are fluffy, faintly streaked juveniles, scarcely foreshadowing the suave adults they will become by next spring. The birds stay for about ten minutes, the young spending much of the time preening, wing stretching, and tail fanning. My attention is finally taken from the waxwings by a flock of warblers in the willows below. They move about so actively, rarely pausing for more than a second or two, that I can identify only a few: two immature male yellowthroats, an immature or female Canada, an adult male Wilson's, and an immature blackpoll. As soon as I try to approach, the warblers move away quickly like leaves before a wind.

Two tiger moth caterpillars munch on separate balsam poplar leaves. The larger is over an inch long, thick bodied, and banded alternately in black and brick red. Along the middle of the back the black is startled by irregular splotches of white which are repeated on the sides in pale yellow. Tufts of pale, stiff hairs bristle from the red bands like cactus spines. I am intrigued by the apparently endless variations and colour combinations of insects.

The thick, ear-like folds of several jelly fungi issue from the moss-lined bark of a fallen poplar. They look as if they had been cut from cheddar cheese, but when I touch their smooth surface I feel a layer of slime. As I go deeper into the forest, and shade, and dampness, fungi bloom like archetypal flowers—rose-red, orange, ochre, yellow, white; thrusting up from the past through the tight layers of dead leaves. In a few days, once the spores are shed, decay sets in; the cap discolours and softens; soon disintegration is complete. I examine some of these older heads. Many contain a score or more of tiny black-headed white grubs, the larvae of fungus gnats which feed on the stalk, gills, and cap; others host only the predatory brown and black larvae of small rove beetles.

I enter a cool depth of spruce at the bottom of the hill along the bog. The sun suddenly seems very far above. Boreal chickadees call huskily. Piping softly, a red-breasted nuthatch flies to the top of a spruce and begins spiralling down. All the way it remains on the trunk as if held by magnetism, although it stretches often to the limit of its short legs to pick little insects from twigs. Before reaching the base of the tree it turns and hitches part-way up again, zigzagging from side to side. Abruptly it flies to the top of the next spruce. The piping is released again, to cease the moment the bird alights.

Occasional crane flies and green lacewings float through the limpid air as though wafted by currents.

The open slope steeps in the sun, lazily reclining, spread wide to the welcome heat like a sunbather on a beach. At the bottom a single aspen grove casts an oval of shadow on a grass meadow pocked with bison wallows. Beyond, the grass becomes clumped sedges and round-topped willow thickets that enclose the blue eye of a pond. Clay-coloured sparrows *tick* in the shrubs, down out of sight. Two goldfinches rollick by overhead, briefly filling my ears with their musical calls. The stillness returns. In all that broad vista dragonflies are the largest moving forms. I sit for many minutes.

A patch of tan appears in the sedges. Then I see the whole coyote. It is carrying something, I am not sure what, in its jaws. A few chewing motions, followed by swallowing, and the coyote walks on slowly, intent on the ground ahead. It is hunting voles. I begin to creep down the slope along the trees, advancing only when the coyote turns away or pounces, and hoping the breeze will not eddy my scent to it. After a quarter-hour I am only fifty feet away. Although I have heard coyotes chorusing almost daily for many months, until this moment I have seen little other than glimpses of the animals themselves. My opportunity for closeness now rests not so much on my stalking skill, but on the coyote being a pup who has not yet developed the extreme wariness of the adult. Not all its pounces yield a vole, yet it appears well fed. The long, slender muzzle gives its head a deer-like delicacy, not out of balance with the rest, for the legs too are long and slim. From a distance the animal had appeared uniformly light, but now I can see detail. The thick fur is grizzled, black hairs mixed in on the upper face, along the back and at the tip of the bushy tail. The tan is purest on the legs and the back of the ears, being blended with greyish elsewhere except the underparts which grade to creamy white. The coyote glances at me occasionally, but shows neither alarm nor further interest. It seems to locate the voles first by sound, walking forward slowly with muzzle dipped, and ears angled far forward. It stops, tilts its head sideways a time or two, then pounces suddenly with stiff front legs. When successful the coyote grabs the vole in its jaws and still standing, quickly eats it.

A wail comes so faintly I am not sure I hear it at all. But almost immediately the coyote throws up its head and replies, the long, high note breaking into a tumble of yips. Then it turns and trots off gracefully in the direction of the call.

I climb the slope again, aware now of the singing grasshoppers I do not see as well as those which soar up in clicking flight, dark hind wings spread open like sails. Nine immature chipping sparrows flush from a patch of buckbrush and flow across the slope to the far side. Several robins and a blue jay fly from choke cherries, the fruits now ripened to obsidian spheres.

Colonies of black aphids soil the stems of creeping thistles just below the clustered pink flower heads. At three, adult ladybird beetles are steadily eating into the soft bodies, and at one, a fat hover fly larva is esconced in the middle of the colony. Ants are in attendance, but make no effort to drive away the beetles, which exude a strong repellent when disturbed.

The flowers still attract a few butterflies. Bumblebees come and go, some with bristles worn down to the smooth chitin and wings frayed; others appear quite fresh. Yet all except the new queens will soon succumb to frost.

I spend the early evening on a wooded headland on the south side of the lake. The breeze has calmed and the lowering sun is warm.

During the summer the lake has dropped about a foot, exposing here a strip of bare mud. From somewhere—I never seem to see sandpipers until they are on the point of landing—a solitary sandpiper alights on the mud. It settles its wings and immediately wades into the water, sometimes deep enough to wet its white belly. It dips its beak repeatedly as it walks along the shore, presumably taking the aquatic insects which still throng the water. A few times it makes short runs, neck outstretched, beak almost skimming the water, and I imagine little sticklebacks or minnows darting away from the overreaching ogre. But the sandpiper does not persist and soon resumes the more rewarding insect dipping. The bird feeds thus for about twenty minutes, silently, its only other action being the dignified bowing that accompanies walking. Then it stands in shallow water facing out over the lake, motionless, head drawn down, white-ringed eyes half closed.

With raucous, shattering croaks two great blue herons fly low over the trees, so near that they seem to shut out the whole sky and pull everything under their dark, rustling cloaks of wings. The sandpiper flies instantly, speeding from under the gloom of the herons. Then the giant birds are gone and the sky of day restored.

To my surprise in a few minutes the sandpiper comes back, at least I assume it is the same one. It feeds for a few minutes. Then, instead of resting when it pauses, it begins some sort of dance. Standing in water half-way up its legs, but without changing position, it raises and lowers one leg then the other rhythmically at the same time as it vibrates its tail laterally. In less than a minute this movement ceases and is followed by bathing. The sandpiper dips its beak and flicks water upwards and over its head and neck; lowers into the water, opens both wings, and jerks them back alternately to the sides with a cupping action that splashes water over its back. Then it straightens its legs and commences preening. Only the beak tip is used to probe back, wing coverts and rump, but on the breast and underparts almost the whole beak is slid into the plumage with a sideways scissoring action. Finally, oil is spread by sliding the beak and head over the surface of the feathers.

At this point something startles the sandpiper again. It wings rapidly across the bay and does not return.

The sun slips behind the trees and I feel the cool almost at once. Ducks and coots in shadow are dark blobs, those still in sunlight are cast in relief. Even the most distant gulls shine white like planets in a galaxy. A small flock of cedar waxwings passes over. They seem in a hurry and their calling is urgent.

The sun is down. Cirrus glitter as gold filaments floating high in the blue, above the molten copper of the west and the sash of rose banding the east.

Ten minutes after sunset dusk darkens the spruce clustered at the shore. Soundlessly the pale apparition of a great horned owl glides down from a spruce where it has drowsed throughout its night of the day. The owl alights on a bent over aspen, folds its wings, and shrinks by half. It has perched facing me. Now it bobs up and down on its legs, at the same time bending forward and turning its head from side to side. I have to smile at this seemingly comic performance, but the droll actions are probably just the owl's way of limbering its muscles after hours of immobility. Perhaps many birds do this on awakening, although I have never observed it before. The owl stills for a few moments, then spreads its wings and is airborne, flying low, wings reaching deeply. Soundlessly it flies, and soundlessly disappears, following the shore.

The clouds tint pink; the west pales; the east goes grey, night hovering just above the horizon. Dragonflies stay aloft. Coyotes chorus, one group off to my left, a second behind and much farther away.

Slowly darkness claims the forest. The clouds have dimmed and fused with the twilight sky. Now only the occasional dragonfly twitches through air heavy with cold; but two little brown bats flutter above the forest, nervously animated black silhouettes. The purity of a white-throated sparrow ascends once, twice; no more. Summer is laid to rest.

A half day of rain from heavy, weeping clouds; then clearing to a fiery sunset—yesterday. This morning grasses, sedges, flowers, and leaves are all crisp-white frosted; rain pools are skinned with ice.

A flock of a dozen juncoes, adults outnumbering immatures, feeds on the ground near the house. With feathers fluffed in the cold they are rotund dumplings of birds, hopping about scratching, dry call notes pattering.

During the past week more and more coot have moved onto the lake. Along the west side I count a minimum of three hundred and fifty. Their appetites seem insatiable as they ceaselessly dive and retrieve lengths of pondweed which they then peck and shake. Widgeons, mallards, and a few gadwalls, sex and age still confusing, crowd among the coots. Often when a coot surfaces three or four widgeons converge on it avidly grabbing at the pondweed which they also relish, but being unable to dive, frequently cannot reach. The coots do not seem to resent this piracy and if pressed too closely merely abandon the strand and dive for another. As I watch, the loud lament of Canada geese swells above the croaks and mutters of the waterfowl on the lake and soon is completely dominant. Then I find the birds, flying south over the middle of the lake, an uneven wedge of thirty-seven. They must be rested and well fed, for they show no inclination to come down.

Two spotted sandpipers in plain winter dress work along the shore flying lightly from one log foothold to the next, pausing only briefly to tip the water for food.

By late morning it is warm again everywhere but in deep, sunless hollows and within stands of spruce. Twice I step over rust and black woolly bear caterpillars humping rapidly across the trail in their search for a winter retreat. Bumblebees, wasps, and hover flies hum in the remaining vetches, asters, goldenrods, hawkweeds, dandelions, and thistles. A tattered and faded fritillary meanders over a meadow without stopping at flowers.

For some time the forest is almost silent. The nasal bleating of a white-breasted nuthatch, the derisive screams of two blue jays, and the *pink*-ing of a hairy woodpecker are all isolated events haloed by inert intervals during which I become aware of yellowed leaves, glowing rosehips, and bunches of sickle, maroon dogbane pods.

Then I descend to a sedge meadow where the bordering shrubs move with small birds. Before I reach them a grey-cheeked thrush flies from the ground, stops in a sapling, wings on through the trees. The flock beyond comprises Tennessee, orange-crowned, Wilson's, myrtle and Canada warblers, most immatures and females. Two least flycatchers snap at insects from higher perches.

Under the shrubs the ground is open where marsh marigold leaves have withered away. Here, after a few moments searching, I locate three white-throated sparrows, three immature white-crowned sparrows, a waterthrush, and an ovenbird. So alike are the birds' contact notes that they could all come from one species, with the exception of the flycatchers' sharp *kwut*'s.

I step out of the trees at the top of a slope and am embraced by summer. The whole shrub-fleeced incline is an island of a season almost banished. Green predominates; flowers are blooming; insects seem as numerous and varied as ever. I take off my jacket, push up the sleeves of my sweater and lie back to be cooked. I concentrate on the heat of my bare skin, trying to imprint the sensation indelibly so I can recall it physically during winter days of absolute cold. For a time I am aware only of my hot arms and face and the blood pulsing through the expanded vessels. I seem to be swelling like a dry sponge soaking up water. But when perspiration trickles, I retreat to the forest, half in shade.

Robins stream over the trees behind, and down the slope in a straggling, fast-flying flock, thirty-six in total. They call frequently to one another, *sreeeee..kuk,kuk,kuk*, a command to keep up with the group. At the bottom they roll up and over the tree crowns like swimmers breasting waves.

High overhead a red-tailed hawk soars in slow, wide circles. Gradually it trends southeast and the circles flatten to elipses.

Birds move along the edge of the slope, keeping to the saplings and shrubs. I move slowly closer. At first I see only the red-orange and yellow flashes of redstarts, the one male orange, the females and immatures yellow. They dominate the flock with their bright hues and vivacious activity, but I also find two young orange-crowneds, three Tennessees, and a black-and-white warbler. A red-eyed vireo moves more sedately above the warblers, picking off the last cherries and saskatoons. A *pink* note leads me to a male downy woodpecker. He hitches about on even the smallest twigs, pecking and probing every irregularity with intense preoccupation. He has little in common with the other birds, yet seems to be moving with them, at least temporarily. This woodpecker is partially migratory in that some individuals journey far south of their summer ranges for the winter, while others elect to stay behind. Like a band of revelers, ten black-capped chickadees dance up from the forest. Sprightly, gay, they flit about from perch to perch, never still, chuckling and twittering continuously. I expect them to move on soon, but when I leave they are still with the warblers, having adjusted to their slower drift. Low in the shrubs and on the ground immature white-crowned sparrows hop about, and near them I once glimpse a Swainson's thrush.

I walk on through forest. Near a bog I begin finding mushrooms wedged into the forks of branches several feet off the ground, first in aspen, then in the spruces. They are all stalked, short-lived species, and of medium size. Red squirrels are the harvesters and when the fungi are dry the squirrels will add them to their store of winter food. But they pass by the little globes of the now mature puffballs which have darkened to the same tan as the fallen needles and are filled with dust-dry spores. They also show slight interest in bracket fungi which become dry and woody with age. On an old fallen spruce, so long dead that the exposed underwood has turned black from decay, cluster the little, half-moon shelves of

a polypore. Since this species is perennial each year's growth is marked by an annulus, the colour progressively darkening from bright orange at the rim to deep russet at the old core. A red-breasted nuthatch pipes in the heart of the bog, a thin, metallic sound that is like a sudden chill.

From the top of the valley I look down across a broad expanse of sedge that fills the bottom. A mature bull moose steps out of the willows at the far side. Although weighing less than a bull bison, this one appears more massive with his heavy shoulders, thick neck, and flaring upsweep of palmate antlers. The now hardened, clean bone gleams palely in the sunlight, ready for the rut. For perhaps a minute the bull stands alert looking across the open; then browses, stripping mouthfuls of yellowing leaves from the willows. I turn away. A chipmunk sits up, front paws across its white chest, and cheeks bulging with seeds. Then it flicks its tail and is gone.

Along the lakeshore whenever I near a beaver lodge I come upon freshly hewn aspen stumps, sometimes neatly conical, but more often skewed to one side. The ivory wood and accumulated chips are startling, as though I had suddenly stumbled upon a skeleton. The downed trees, also with conical ends, lie beside the stumps. When cutting through a large trunk a beaver begins by gouging a notch, then shifts and cuts inward from below, always using the lower incisors to slice, the upper to provide leverage. Many of the chips are thick, over an inch wide, and almost two inches long. Occasionally a tree has not fallen but leans entangled; however, most of those lying on the ground have been limbed and great patches of bark gnawed off as well. The branches have been dragged to the water and sunk into the bottom near the lodge by the weight of more branches above. Over the next few weeks the pile will be further increased to assure food for the entire winter.

I walk the length of the point shortly before sunset. The cattails are empty. A few red-necked grebes bray intermittently, briefly and without vehemence. But ducks are everywhere: mallards, widgeons, pintails, shovelers, gadwalls, and teal close in and broadcast across shallow, weedy bays; scaups, goldeneyes, and buffleheads together with a sprinkling of white-winged scoters, ringnecks, redheads, and canvasbacks farther out. Detachments in flight are dark nebulae in the sky—some coming in for the night, some fleeing the hunters' popping shotguns just beyond the boundary, some launching into the darkness and on southward. Gulls are a minority, but disproportionately visible, like flecks of white paint on a dark canvas.

Dragonflies still ply the air, while thousands more await a new summer as aquatic nymphs in every pond and lake.

Near the tip of the point I come upon a cluster of new aspen stumps and the initiation of a subcycle within the larger ecosystem. Next summer grasses, flowers, and shrubs will proliferate in this sun-drenched clearing and in consequence attract and sustain a different variety of insects; appeal to "edge" birds such as chipping sparrows, yellow warblers, redstarts, catbirds, cedar waxwings and kingbirds; yield more rose hips, currants, gooseberries, raspberries, saskatoons, cherries, hazelnuts, and bush cranberries, flower and grass seeds, for birds as well as chipmunks and mice; produce more winter browse; and provide a substrate, in the dead stumps and abandoned trunks, for wood-boring insects, bacteria, fungi, and mosses. Aspen saplings will soon sprout from the living roots like a thicket of cane, and in a decade shade from the young trees will suppress the shrubs and herbs again. However, as long as the beaver are present the disturbance will be repeated annually, only the site shifting as they seek out mature trees.

Abstracted in this train of thought I have been still for some minutes. Slowly I become aware of an animal nearby, shuffling through the leaves. The beaver and I see each other at the same moment. It seems to peer at me myopically, swinging its flat, squared head a little and twitching its nose to catch my scent. But these beavers are accustomed to humans and it does not move until I do. Then it pivots and starts to bustle away; however, as I step off toward the house it stops. I leave it to its nocturnal industry.

Another calm, clear night with the thermometer reading 27 degrees

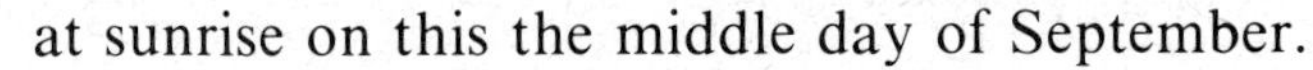

at sunrise on this the middle day of September.

The lake is strangely empty of waterfowl. The majority must have left during the night, the weather in their favour. But soon a swirl in the ebbing tide will bring new southbound contingents, and more after them, until the final ripple retreats to leave the lake as clean as a wave-swept beach.

As the day progresses and I circle more of the lakeshore I see a number of waterbirds, although they are nowhere numerous. In one bay an immature pied-billed grebe dives at the edge of twenty-five busy coots. A male belted kingfisher rattles from a dead snag at the edge of an island. Then he is silent, staring into the water. My next impression is the confusion of white water from his dive, then his emergence with a stickleback crossways in his beak. He returns to his perch, beats the fish against the branch, and gulps it hurriedly. Just as he settles to continue his watch a second kingfisher swoops in from behind and both birds career off stuttering excitedly.

Fifty Canada geese approach on a southeasterly course, calling continuously. Over the lake itself they come down lower and their flight conversation changes to one of discussion. I think they will land; however, they decide not to and beat on.

Half an hour later I am on the trail along the north shore and pick up the distant *wah*-ing of white-fronted geese. As they near the volume swells and the pitch rises until the sound is almost a cackle. Then the big birds are overhead and I catch glimpses of them through gaps in the foliage. With their brown instead of black-stockinged necks they are at once distinct from Canada's. I estimate the flock to number between fifty-five and sixty, and it too hurries on bearing southeast, perhaps headed for a stopover on Beaverhill Lake. Thousands of migrating waterfowl, both spring and autumn, traditionally mingle to rest and feed at this very large, shallow, and protected lake.

An immature red-necked grebe, still faintly striped on the head, swims well off-shore. It is silent and I cannot find any others on this part of the lake, although two horned grebes float not far away. I hear a kingfisher again, probably one of those I saw earlier on the south side.

From shore I count seven buffleheads, six drakes showing the first signs of fresh nuptial plumage and one hen or immature; seven common goldeneyes, all resembling hens; eight mallards, among which I distinguish two drakes by voice alone; a gadwall who comes out onto a log and begins to preen; and about thirty adult ring-billed gulls swimming and flying out towards the centre of the lake.

Two small flocks of Canada geese wing over, one near, the other off toward the west shoreline. Three wedges of sandhill cranes, each of twenty to thirty-five, follow soon after, but much higher so that their bugling arrives thin and tenuous. The calls of geese and cranes seem now to lack the eagerness that announced these migrants' northward trek. The difference may just be in my own emotional state, yet the birds' physiology has also changed and the autumn return is usually more leisurely than the spring drive to reach the breeding grounds.

At intervals, where I found them during the summer, I come upon song sparrows in the shrubs, but usually solitary birds. An adult red-tailed hawk falls from a tree and flaps across the lake, silent and unhurried.

Landward, lost in the trees, I can hear two or three blue jays together, ripping the air with their cross-cut screeches. Four chickadees come close chattering and occasionally singing. Good weather stimulates these sprites to sing at any time of the year, and I have even heard their whistled song on mild, sunny January days.

At midday the sun is almost hot; the air and ground warm. Insects are active once more, but the two frosts have decimated their numbers. I send only one damselfly from its basking site, although dragonflies are quite common. A few bumblebees and yellowjackets visit the lingering asters, thistles, and beggarticks; no hornets. But a red admiral flies strongly along the trail, vivid and unblemished, pausing briefly to sample one blossom then another. Ants wander the bare earth and meet like a crowd at a market on the dome of their nest. I catch the scuttle of ground-toned wolf spiders, one trailing her whitish, bustle-like egg case. I am surprised by the abundance of adult grasshoppers and their occasional stridulation—life lived to the full until the last moment.

I stand on the high bank just before the trail bends inland. Far enough off not to take alarm an adult great blue heron stands a few feet out in the water, its neck half folded. It seems absorbed in contemplation, or vacuous, but I doubt either interpretation. The success of its usual hunting technique depends on long immobility.

Four crows fly southward over the lake, silent, dead black against the vibrant blue sky.

Warblers again occupy the willows: a few Tennessees, orange-crowneds and Wilson's, one each of mourning, magnolia, and yellowthroat. A flicker crosses the open of the bay briefly distracting me with its ostentatious, bounding flight and flashing white rump.

I shiver in the raw chill of the spruce by the bog. The sun never really reaches to the bottom of this north slope. In winter it is too low in the sky, while in summer most of its light is blocked out by foliage. The frost has melted here too, but it is so much cooler than the sunny lakeshore that autumn seems already to be gone. I walk past the ghosts of oak ferns, horsetails, mayflowers, fairy bells, and white violets. Out in the bog itself the cloudberry and three-leaved Solomon's seal are fading and withering, revealing the crimson bog cranberries and the tips of Sphagnum reddening like nose and ears in the cold.

The white spruce thins; the crest of the hill descends; shafts of sunlight brighten the ground and probe into the bog. Several juncoes infiltrate the thick shrubbery, but are far from secretive, hopping over the ground, scratching in the leaves, flying up to low branches with suddenly white-edged tails, down to the ground again, chasing briefly as one infringes on another's space, constantly *tick-tick*-ing. One junco catches a small moth flying over its head, rolls the body between its mandibles as it would a seed, then swallows it whole. A white-throated sparrow and a waterthrush feed quietly on the damp ground of the moat.

With burred, interrupted notes a little flock of siskins rushes into the top of a spruce. The birds seem to disappear and it takes binoculars to distinguish them clinging to the twigs and attempting to peck into the cones, most of which are still closed. In my search for the siskins I find, lower down, a little group of myrtle, black-throated green and blackpoll warblers working quietly in the spruce as though confined to a room. Suddenly a red-breasted nuthatch calls, also from the spruce. A blue jay planes down from another spruce, holding a cone in its beak; lands on a dead aspen branch; drops down into the undergrowth. Soon it emerges, its beak empty, and flies through the trees.

I wait for several minutes hoping the juncoes will move away from the trail, but they seem settled, and eventually I must walk through them. Some dart into the bog, others up into the white spruce with a confusion of flicking white and a flurry of hard *tsick*'s. I look back from the top of the hill. The juncoes have reassembled where they were and are feeding again.

Goldfinches fly overhead, invisible, their rich flight notes transiently filling a habitat they will never enter.

I come down to one of the seeps. Here a wood frog sits in a pool of sunshine, breathing shallowly but rapidly. It must soon find a burrowing place in deep, soft leaf mold or under a rotting stump. Downstream a robin stands drinking from the trickle of water. Its breast feathers are pale-edged and a few of the lower ones spotted with black, indicating immaturity. Then it flies up into an alder, shakes, sleeks its plumage and flies again calling softly, *kuk . . kuk . . kuk.*

Despite many coloured, withered, and fading leaves, green still prevails. Seasonal transformations are gradual, some plants and animals responding early to the first hints of change, others reacting tardily or in ways not immediately apparent. Most mushrooms have suffered from the frost and are liquifying like plastic but the hard bracket forms and the dry sacks of the puffballs will look much the same next spring as they do now.

A red squirrel jerks up a smooth aspen, the tailed brown hazelnut in its mouth not hampering its imprecation of me. I have interrupted its harvesting and perhaps will spy on it to raid its cache of winter food. The squirrel cannot conceive of an indifferent observer.

Several bunchberries have been ravished. Fragments of the scarlet fruits are scattered about and only the seeds taken. This may be the squirrel's doing, but more probably the small seeds have been eaten by a chipmunk or mouse.

I hear a white-breasted nuthatch and find it inspecting aspen trunks minutely—pausing to pick at each crevice and knot, then moving on in its characteristic abrupt manner, a meticulous, black-capped grey and white bird.

I leave them behind. An ichneumon fly lands on a hazel leaf at eye level. Her orange-brown body alone is nearly a half-inch long, yet from the tip of her abdomen the thin, black ovipositor extends back over twice that. Ichneumons with such lengthy drills lay their parasitic eggs in the tunnels of wood-boring insect larvae. I move and she flies away continuing her quest for riddled trees.

The air is calm, but now and then an aspen leaf lets go and twirls down slowly, sometimes catching in the shrubs, sometimes falling all the way to the ground.

A male hairy woodpecker hitches up a dying aspen infected with heart rot, pecking here and there, but without finding any prospects worth the excavation. He calls intermittently, then is off to another tree where he continues the same testing search.

Along the top of the last ridge, with spruce and birch on the slope, I encounter another little gathering of myrtle warblers and just past them more juncoes. To some extent the daytime associations of small migrating songbirds are fortuitous, dependent on coincidental meetings. On the other hand, mixed flocks of closely related species are probably more cohesive and a particular group may remain intact for days, or even weeks.

A painted lady butterfly rests on the broad dais of a cow parsnip leaf, sunning its black, orange and white wings. Another flock of Canada geese trails overhead, honking, and bearing southeast like the earlier ones.

A dark, rod-like object several inches long lies on the road ahead. I stop, walk up to it, and find a tiger salamander, tubular like a cigar, with a blunt nose and long tapering tail. The short legs appear totally inadequate. This salamander's name derives from the irregular vertical bars of greenish-yellow on the sides, legs, and tail which break the background blackish-olive. The salamander is motionless, but when I pick it up it croaks once and squirms in my hand, the long-toed feet pushing with unexpected strength. It must have started to cross, then found the warmth of the pave-

ment alluring. In the interests of its survival I carry it to the far side of the road and put it down just inside the forest where it can find a place to hibernate. I do not know whether these salamanders are genuinely rare in the park, or whether their secretive habits and silence just make them seem so. Like frogs they mate and lay their eggs in ponds, but the adults soon return to damp forest and there spend daylight sequestered under logs or down in the leaves where the humidity is high. I have seen countless tadpoles, but never any of the elongate, greenish salamander larvae which, except for external gills and colour, resemble the adults closely. Indeed, I have seen adults only a few times, usually in spring and autumn when the cooler, damper air enables them to move around with greater freedom.

I find more ducks, coots, and horned, eared, and red-necked grebes on Tawayik Lake, but even here they are not abundant. South near the narrows I pick out thirty Canada geese on the water, gigantic beside the ducks. Bison are too numerous for comfort and I never go far from big trees; however, they are more concerned with grazing than with me. The time for green grass is short. The rut is over and bulls again graze side by side with scarcely a glance; indeed now they prefer the company of their own sex. Most of the calves are big, spike-horned, slightly humped, and as dark as the adults. All are becoming shaggy-coated, the lengthening hair obscuring the ripple of moving muscles.

Magpies fly about in chattering groups, apparently drawn to

the bison. From time to time one alights on the back of a bison for a short time, although at too great a distance for me to discern what any are doing.

Robins pour out of the trees and across the meadow, then over the lake in a ragged flock of forty-one.

Ungrazed clumps of upland grasses are drying to tawny hay. All the goldenrods have gone to seed, and most of the thistles. Crimson-leaved fireweeds, their pods burst, are wreathed in clouds of pale smoke. Only the yellow-disced asters still weave nets of mauve, and a few dandelions stamp yellow circles on broken ground and the margins of wallows. The last beggarticks glow like suns among the cattails, bulrushes, and sedges dying back from the tips of their leaves.

On the barren dome of a field ant nest many of the inhabitants run about, entering holes, emerging again, running left, running right, touching antennae with another resident, rushing down the slope and disappearing into the towering grass jungle. A few bumblebees, an alfalfa butterfly, and a Milbert's tortoiseshell visit the asters. The only flies I see are dark, hairy muscids, a few hoverflies, and scintillating bluebottles. Grasshoppers arc up from the grass like sputtering fireworks, then plane down on sail-wide wings; others I cannot see float their rasping songs on the air.

As the late afternoon shadows extend dark paths across the meadows I hear from a long way off a bull elk's squealing grunt. It comes again and again—challenge, advertisement, and appeal. I walk on and soon pass battered shrubs and saplings where an elk or moose has thrashed his antlers, to clean them of velvet or perhaps to relieve pent-up emotion. The torn leaves and bark hang in drying shreds.

Later at home, with a quarter moon in the darkened east, the hooting of a great horned owl drifts over the water from the north shore. When it stops I hear only the coots' protesting notes as they feed relentlessly.

A long, dreary day of cold, misting drizzle and raw, driving wind from the north; bottomless clouds a grey gloom above the trees, with neither beginning nor end.

A song sparrow flits through a willow; juncoes are in little groups along the forest edge and road, and with one of these are two chickadees and several myrtle warblers, all subdued; a robin alone calls querulously. These are the only land birds.

Mallards, widgeons, and coots by the hundred huddle near lee shores, although farther out goldeneyes, buffleheads, scoters, grebes, and a winter-plumaged loon bob unconcernedly, dive and preen. A few ring-billed gulls ride high among them, like deep-hulled white yachts.

With dusk the whish and rustle of beating wings are four sandhill cranes winging scarcely higher than the trees, slender necks and trailing legs outstretched, calls deep and resonant, and spine-tingling now that they are so close. They speed on without hesitation, perhaps trying to outfly the weather or looking for an open landfall.

Dawn on the twenty-first of September. The ground, the drooping grasses and flowers, leafy shrubs, and branches are all covered thinly with snow as if a white fungus had grown overnight. Ice seals shallow standing water.

The sun arcs into a flawless sky on which compact detachments of ducks, and arrowheads of geese and cranes are darkly silhouetted with the precision of a silkscreen print. From the house, looking east and west of the point I estimate at least a thousand waterfowl: coots, mallards, widgeons, shovelers, gadwalls, pintails, and a few teal inshore; canvasbacks, redheads, goldeneyes, buffleheads, white-winged scoters, and scaups elsewhere. I can occasionally find a recognizeable drake. Horned and eared grebes are sprinkled among the diving ducks, and over toward the north shore of Long Island I find twenty-one red-necked grebes, nearly all adults. With them are four western grebes, larger, slimly elegant, their white cheeks and necks shining in the sun. On a third sweep over the lake with the telescope I locate a common loon, three female or immature common mergansers, six elfin Bonaparte's gulls, and near the last, three northern phalaropes. The phalaropes are rarities to be seen only during their migrations between breeding grounds on the Arctic coast and offshore ocean wintering areas.

The sun is warm and where it strikes, the snow soon begins to soften, the white crystals blurring, then clearing and rounding to drops of liquid. By early afternoon snow lingers only in permanent shade.

A loose flock of forty-eight crows straggles just above the trees on the far side of a meadow. At any distance these birds always look matte black; only when they are very close is sheen on their feathers apparent. I walk across the grass and from a lone aspen a red-tailed hawk flings away, staying in sight barely long enough for me to see the darkly barred tail of the immature. Then in an aspen grove I encounter juncoes together with white-throated and immature white-crowned sparrows. I walk through the centre of the grove and a grey-cheeked thrush flies up and between the widely spaced boles.

I look down on a sedge meadow where a coyote is vole hunting. I think the animal a pup until it suddenly stiffens, glances my way, then streaks off low through the sedges to the nearest trees. A shift in the breeze may have brought it my scent, or it may have just felt me watching it. I wish I could wear the mask of a bison, or moose; any animal but man.

The sun sets, turning the waterfowl to dark cutouts on the painted lake. A robin flies past me calling softly, unafraid of the coming night. Coyotes unleash their wavering cries to flow through the forest, across the lake, into the sky, ultimately to oblivion.

Summer has returned. All the snow has melted. The gentle southerly breeze whispers sweetly through the leaves of times past, procrastinating.

But in the four days since the storm, change has accelerated.

The ubiquitous green withdraws, and like a curtain being pulled back hidden pigments are revealed. Balsam poplars stand to attention like bronze-armoured knights; aspen leaves are burnished gold coins in the sun; hazels, saskatoons, and birches reflect the aspens' gold. Other shrubs flame orange and scarlet. Red-osiers, bush-cranberries, and bunchberries are glowing crimson splashes. In the bogs larches are a race apart, segregated by colour as their deciduous needles turn yellow. Only alders and buffaloberries keep company with the evergreens—most of their leaves will drop as they are, or change to dull, leather-brown.

The last seeds have ripened: slender chestnut catkins on birch; stubby dark cones on alder, opening now to free the seeds; tan pendants on spruce, from which a few winged seeds float free and spin in the breeze.

This morning several crows assemble in the trees on the point. They slip from perch to perch, flap their wings, caw and croak, all preparatory to their eventual flight together, but west, not south. It will be a few weeks yet before winter drives them from the country.

Waterfowl are still thick on the lake, the same species as a few days ago except for the teal and western grebes who have gone. On the far shore of the bay behind the house a statuesque great blue heron waits at the edge of the water, its attenuate reflection blending with those of the poplar trunks all around.

I walk to the end of the point. The feed bed beside the lodge is delineated by a bristling beard of twig tips rising out of the water. Where the shore vegetation has been trampled flat by the comings and goings of the beavers I find a few mink scats close together. Two contain only mammal fur and bone fragments, one nothing but water beetle skeletons, and the fourth is entirely shrimp remains. All the scats are about the same size, and may well be from the same individual.

A tree sparrow, the first of the autumn, is with a flock of juncoes and white-throated sparrows. And in the willows edging a sedge flat are the next to last warblers: a few myrtles accompanied by three orange-crowneds, an adult male Wilson's, an immature mourning, and a female palm. In ecstasy the one male myrtle warbler sings repeatedly, a light twittering ditty quite unlike the spring territorial song. Midges, fungus gnats, and other small insects have responded to the heat and the warblers are catching them.

By early afternoon other insects appear: occasional mourning cloak, angle-wing and Compton tortoiseshell butterflies; some little, bright tan moths; a scattering of muscid flies and bluebottles; some small ichneumons; and a few dragonflies. A flicker bounds up from a meadow and I find where it must have been licking ants from a nest.

I follow the stream valley south from Astotin Lake. A female kestrel hovers over a sedge flat, a quivering, autumn-brown leaf. Then she drops, submerging in the sedges; and surfs up a few moments later with empty claws. She flies a little distance and hovers once more. After her second plunge she rises with something dark and angles off to the trees where she can perch and eat. Her wings beat deeply and strongly, but I am left with a feeling of reserve power unused. At this moment another hawk flies from a dead tree stub and out low over the ground. Its white rump and long wings immediately mark it a marsh hawk. Its flight is unique, being buoyant, and the wings lifted well above the back. As it courses down the valley it veers slightly from side to side, a graceful, ballet-like progression. In less than a minute it wavers out of sight around a bend.

Elk bugling thins in the air. Arriving from a very great distance the summons lacks authority despite repetition.

I walk along the valley all morning. However, after the hawks and elk, except for overhead flights of ducks and geese and a few small groups of pine siskins, the place seems deserted.

An empty sky has no perceptible depth, but for the past few hours cumulus have been growing and massing, puffing up higher and higher into solid, heavy-based mountains. Yet blue and sun are not completely vanquished; bright shafts beam through openings whitening small detached clouds and firing the vivid trees. As the clouds drift on, sun and shadow exchange places in a shifting mosaic of meadow and forest. When I return to the lake in mid-afternoon the mirror calm inshore doubles the cloud. Suddenly the earth seems to be a flat, thin crust mossed with a dwarf forest, suspended in the middle of space and diminished to insignificance. The shoreline is the limit of the world.

A ruffed grouse drums; a minute later drums a second time, clearly, unmistakeably. Unseasonally warm weather like this often revives singing or its equivalent. This morning it was the myrtle warbler; now the ruffed grouse; and shortly after, I trace a repeated rambling warble to a male purple finch high in an aspen.

At the edge of the water a robin drinks, leaning far over from its log footing, and once or twice flicking its wings for balance. Two song sparrows flit about in the willows, calling intermittently, this time a harsh warning *chenk* note. Somewhere in the distance a cow moose yearns for a mate with a prolonged nasal bawl. I hear her again several times, but no audible response from a bull.

The clouds are still heavy at sunset, and only their bases flare crimson, fade to pink, dull to iron-grey. Then it is night.

October

A week into October this morning, and I hear Canada geese before I rise and dress. Outside it is dull under an overcast sky, and frosty. A tough breeze whips out of the west winnowing great flocks of leaves, so that the whole forest seems to be flying away. Where the wind reaches across the lake the water is rough, and dark, and remote. Perhaps it will snow, although the clouds are tightly rolled, without the frayed edges that presage imminent snow or rain.

I take the telescope to the end of the point. The ducks are fewer now, only two to three hundred scattered over the breadth of the lake from the northwest cove to the southeastern concentration of islands. As usual most of the coots, mallards, widgeons, and pintails feed in the shallows, with the diving ducks beyond. Nuptial-plumaged drakes now account for perhaps five percent of the total. There are a few horned grebes and ring-billed gulls.

As I start to ready the telescope for carrying, a large almost black bird flies out over the water from the north shore—an immature bald eagle? But no, when I focus the hawk has the proportions of a red-tail in spite of its dark body and wing linings. The greyish-white tail and flight feathers are edged with black, the underparts irregularly flecked with white. Formerly known as Harlan's, this hawk has now been relegated to a subspecies of the red-tailed hawk, and by so doing the complexity of this continentally distributed species has been increased. The "typical" red-tail has two dominant colour phases: "dark" (reddish-brown body and wing linings, rufous tail), and the more common "light" (brown above, streaked white below, pale wing linings, and rufous tail); but so does the Harlan's (its light phase is almost white below). Within each subspecies the two phases cross to produce a variety of intermediates, and I suppose it is further possible that the two subspecies may occasionally form mixed pairs in the northwest where their summer ranges overlap.

Without sunlight the foliage loses its brilliance and the colours themselves seem to retreat. I walk through gaily draped halls, but have to reach out and feel the fabric to establish contact.

During the morning I encounter several junco and tree sparrow flocks, but all are widely separated and never number more than twenty birds each, more usually eight to a dozen. It seems strange that there should be so much empty yet similar habitat between these groups. They may represent amalgamated families, mostly of immatures at this stage of migration, from a local area, and not until they near their main wintering region will these tributary streams join to form the ponds and lakes that flood southern fields and thickets.

A blue jay hops deliberately over the leafy ground its head cocked to look down. Periodically it stops and turns over a leaf or two searching for hazelnuts and other fallen fruit, perhaps also insects. A second jay nearby is doing the same. Both birds are silent, the only sound being the dry rasping of their claws on the dead leaves. Jays and magpies are more interesting when they are quiet, because so often we are aware of them first by voice when they are reacting to an external event rather than revealing some facet of their own behaviour.

This beaver pond, any beaver pond now, is ready for winter. In addition to the feed bed, new sticks and mud have been plastered on the lodge, beating down the nettles that crowned it all summer, while the dam has also been reinforced.

On the water the winds have broken up the duckweed carpet and the remnants have been driven into the cattails. Even as the tiny plants die, whiten, and decay, round bulbets sink into the bottom mud. Next spring they will surface, germinate and again proliferate endlessly.

Five mallards, two widgeons, two pintails, and a few coots feed around the shoreline. Toward the centre a black and white drake bufflehead and three indeterminate scaups pop up and down on brief dives like floats being pulled under then suddenly released.

As I walk east from the pond the clouds begin to open—patches of clear sky and sweeps of sunshine. The wind worries at the clouds and soon they are scattered like tufts of wool from a sheep's pelt. I soon forget the grey depression of the morning as the leaves glow, and kindle and finally radiate their own light. Through aspen forest I am washed in the golden luminance of liquid air. It alters the tone of my skin, transforms the smooth trunks to pale lemon and saturates the fallen leaves. A magpie stands on the ground tugging at the shreds of some long-dead animal. Its white undersides are tinted by the reflected yellow. Above, its irridescent wings and banner tail shimmer from aquamarine to jade as it moves. The bird is like a precious stone set on a pillow of gold.

I surf the cresting waves of yellow, orange, vermilion, and crimson, hues unrivaled in intensity, when brushed over square mile upon square mile. They flow through me, become me; I become them.

Exhausted I haul up on a shore of spruce and birch; welcome the cool dark green and subdued amber, the dull russet of the needled ground. Low straggling shrubs are here only highlights in an otherwise sombre tapestry.

Overhead two red-breasted nuthatches are piping; then one flies into a birch. It devotes two or three minutes to an intense inspection of the trunk and some of the larger branches. Clinging tightly, its whole body flattened against the bark, it seems to be a part of the tree. Woodpeckers by contrast, leaning back from their point of attachment give the impression of being in opposition, always ready to attack the tree and make it yield food by force rather than treaty. But this is an exaggeration, for no difference in motivation exists in the birds, yet the analogy does contain a fragment of truth. Nuthatches and creepers pick their food—mainly small insects, their eggs, and spiders—from superficial cracks, whereas most woodpeckers must chisel through solid wood to reach deeply boring insects.

Seeee . . . seeee . . . seeee . . seeee . seeee . . seeee: these slender threads of sound have mingled with dusk-voiced boreal chickadees ever since my arrival. But the birds remain high in the spruce, hidden. I "swish" softly. Slowly they descend, a rung at a time. Then the dusk-plumaged chickadees are close and looking at me curiously. Still the *seeee . . seeee*, which does not seem to be any nearer until I make out two ruby-crowned kinglets a few feet above. They remain among the outer twigs and are difficult to see clearly. I remember longest the white-ringed, wondering eye set in the soft plush of the plain olive-brown head, which gives them a perpetual expression of juvenile innocence. Soon they rise through the spruce again as if buoyed up by a thermal.

I walk down a slope and come out into a yellow cloud of willows. A dark song sparrow flies away through the netted stems. Twigs shiver and in a clump of leaves is an immature male Wilson's warbler. He looks for insects on the leaves and also flutters up to those above. A few midges and fungus gnats are the only insects I see here and the warbler is catching them as they land. A second bird in the same willow proves to be a young female mourning warbler. She has the characteristic greyish hood and interrupted white eye-ring, but is almost white below instead of the usual yellow.

With the rush of a spilling wave siskins pour into the spruce and birch behind. It is easy to single out the young goldfinch from the heavily streaked siskins, even though it is a warm buff, not the brilliant yellow of the summer adult. Its stubby, pinkish beak is a further distinction, for the siskin's is dark and nearly as fine as a warbler's. Except in spring when various insects are eaten, the goldfinch's diet comprises wildflower seeds, whereas the siskin specializes in extracting seeds deep in cones and catkins, although in summer it too is insectivorous.

I walk back slowly through the dazzling afternoon, for this late I never know which will be the last day of genuine warmth. The wax-white snowberries are more like ornaments than fruits now that most leaves have fallen from the wire-thin twigs. Then I find several on the ground, chewed open, and only the seeds eaten —chipmunk, squirrel, or mouse? Some rose hips are shrivelling a little but they will retain their food value all winter. Early in September upper buds on a few young rose canes broke dormancy and now the perfect, bright green leaves seem out of place flanked by the copper, crimson, orange, and yellow of spent summer ones. The way back to the beaver pond is marked by robins in a thicket of dogwood swallowing the spherical white berries as rapidly as they can pick them, a ruffed grouse clucking seconds before it crashes up through the undergrowth scattering leaves, a female pileated woodpecker whacking at a rotten aspen stub, and a pair of hairy woodpeckers working in adjacent trees and *pink*-ing softly to each other.

Standing at the lakeshore in late afternoon, I notice how much the water has cleared. The blooms of pea-green algae are over, the billions of cells dead and sunk to the bottom where decay will release their elements and compounds to support next summer's cycles. Even Astotin Lake with its outlet is never thoroughly flushed by the spring run-off. This results in an irreversible accumulation of sediment, nutrients and organic remains which will in future millenia convert the lake to a marsh.

It is completely dark by eight. I stand out under a panoply of stars. The breeze is a whisper. The only sounds come from coots and ducks until low-flying sandhill cranes flow southward. I follow their passage with my ears, from horizon to horizon. A quarter-hour later another small flock passes. Three times during the rest of the evening, from inside the house, I hear yet more.

A week of crisp, frosted nights and clear, mild days; light breezes. Leaves falling from trees and shrubs even when the air is still; falling all through every day so that now the forest is threadbare. The last aster flowers have withered; green has retreated. In the warmth of the afternoons fungus gnats and a few larger flies take to the air again; once a mourning cloak butterfly. Several times I find a solitary daddy-long-legs crawling on a log. Wolf spiders are still active on the leafy ground. A few ants run about on their domed nests. A small centipede ripples over moss. Bare twigs are strung with the finest of glistening threads, the aerial trails left by young dispersing web spiders. On the warmest day, almost calm, the air itself shines with these same threads drifting as crowds of spiderlings balloon away to somewhere. But after each frost the numbers of insects and spiders decline as some die, and others enter hibernation. Ants are the last to remain active. The ebb tide races.

Still little flocks of juncoes flit through, some mixed with tree sparrows, some all juncoes. This morning I see three groups of tree sparrows only. Alone at the margin of a bog, a fox sparrow hops over the dank ground. It is as large as a white-throated sparrow, but dark chocolate above with heavy splashes of the same on its white breast. In its search for seeds it kicks over leaves with quick backward jerks of both feet together. Beyond, the heath of the bog is sprinkled with lingering, crimson blueberry and bog birch leaves.

The geese are fewer, and the ducks have thinned out even more, although the same species are present as at the beginning of the month.

As I walk to the north shore of Tawayik Lake I meet a band of four chickadees gleaning the poplars. Before I see the water I hear what reminds me of turkey gobbling. I hurry on. There, far down the shore and in the shadow of the trees, eighteen whistling swans

shimmer mirage-like in the waves of warm air rising from the water. But they are real; a mirage does not talk. Six are greyish young of the year, the rest snow-white adults. Several float serenely, long necks stiffly erect; others preen or reach out to gather pondweed at the surface. I watch them for nearly half an hour. Toward the end they feed less eagerly, preen, and rest more. Their conversation rises and falls, now loud now faint, but they move little from where I first saw them. Ducks and a few Canada geese swimming near the swans are dwarfed by comparison.

Reluctantly I leave and walk west, through forest once more open to the sun and reduced to oriental simplicity. From time to time I push through the thin smoke of midge swarms.

Now I hear frequent elk bugling, once the clash of antlers, the rattle of bone against bone. It seems near and I advance slowly and cautiously, but when I reach the little meadow it is empty. I walk on still hearing the bugling. The autumn rut is a severe strain on mature bulls since they have no chance afterwards to replenish their fat reserves, and they often suffer high winter mortality. Nevertheless, because one bull can serve several females in a season, an equal sex ratio is not essential. Among deer species population increase is primarily dependent on the females receiving good nutrition throughout the year. But alone of the North American deer the bull elk maintains a harem during the rut. This creates a spectacle for the lucky observer, but may make this season even more of a trial for the bulls who must be eternally vigilant, collecting new cows, preventing them from straying and repelling rivals. Moose, white-tailed deer, and mule deer seek a single female, stay with her briefly, then look for another. Occasionally males meet at the same female and fight, but overall the tempo is much less violent than for the elk.

I cross the meadow and walk on through dense forest. Before long an elk rises to his feet not far from the trail. He is a young bull, thick-necked and dark maned, but with slender, few-tined antlers. He does not immediately run and his first few steps seem stiff.

I do not see the female downy woodpecker until she flies, a sudden pattern of black and white. A white-breasted nuthatch calls, but too far off the trail to look for with any hope of success.

I come out into another meadow, much larger than the first. A band of cow elk, attended by a big bull, startles away into the trees, galloping and crashing through the undergrowth like a herd of horses. At the far end bison congeal, begin to flow, then ripple away, emptying the meadow. It seems suddenly barren until I find a flock of tree sparrows at the edge picking seeds from the goldenrods, and asters. A troup of four magpies arrives chattering and laughing, leapfrogging along a side of the meadow, and finally crossing it where a lone-standing aspen provides a convenient stepping stone. In a light wind they manage their long tails gracefully, although on alighting almost invariably bow slightly as the tail tips up before subsiding. It is not coincidence that the largely non-migratory jays and magpies possess lengthy tails and short, blunt wings, whereas the far travelling crows and ravens have abbreviated tails and hawk-like wings.

On my way back three evening grosbeaks fly overhead undulating deeply, black and white wings flashing as they ring out their nasal flight calls. This species, like the two crossbills and pine grosbeak, is somewhat unpredictable in movements and nesting in a given year. The park is on the southern edge of the breeding range of all four finches and provides some suitable habitat, mixed-wood and spruce forest, yet in three and a half years I have seen them only from late autumn through winter.

The wind of their flight sings in the wings of seven whistling swans letting down silently for a landing on Tawayik Lake. They are over and gone in a few seconds, but my mind retains the imprint of their splendid white purity.

I have not been aware of autumn willows until now when they release their seeds—white puffy catkins on leafless twigs.

Two siskin flocks feed in the birch along the south shore of Astotin Lake an hour before sunset. In the spruce behind I hear a red-breasted nuthatch and several black-capped chickadees.

A ragged line of crows files across the sun's afterglow at the same time as coyotes herald the arching night. For a few minutes more the island of reeds lies lion-coloured on a darkened plain, then fades.

Later, the starry night in control, the cries of geese sound faintly, and a great horned owl booms, "*hoo-hoo-hoo-hoooo . . . hooo-hooooo.*"

Naked branches rattle as the rough wind tosses them aside. Flurries of dead leaves scuttle over the ground, lift in eddies, fall a second time. The clouds sag unevenly, blur, then the snow drives in curtaining squalls—now blindingly wet, obliterating; now pricking, reduced to a smudgy haze.

I hunch beside the lake, sheltered from the wind. Mallards, widgeons, and coots are close at hand. Out in the open a raft of scaups, goldeneyes, canvasbacks, and redheads rides the leaden water, facing the wind. The dark bodies slide up, slide down as the waves roll under them—the rhythm of the flock is the rhythm of the waves. When I stare fixedly through binoculars I soon lose the identity of the birds and register only the hypnotic motion. Suddenly the whole vision disappears, swallowed up in whiteness.

As the hours slip by the squalls merge and become continuous snow. It hisses through the air; patters on the stiff leaves until they are covered, their hollows and spaces filled; sifts down the grass and sedge stems and is lost; slants into the water, dissolving.

The sibilant night is wrapped in a cocoon of white.

Four days after the storm it warms—a breeze from the south, the sky clearing. The snow softens and slumps wetly—melts away entirely where the sun strikes all day, partially in less exposed places where it creates a reticulated pattern like the pelt of a giraffe, not at all in shade. The small ponds are ringed with a thin shelf of ice that persists. On the third day, with the mid-afternoon shade temperature approaching 50 degrees, a mourning cloak butterfly passes me and alights briefly on a log before flying on.

The mildness holds as October nears its end. Banked clouds this morning soon disperse and the sun liquifies the overnight rime of frost.

The plants I touch are hard or brittle—mummified wildflowers, dried grasses and sedges, spike-needled spruce, leafless twigs. Even the evergreen leaves of bishop's cap, wintergreen, bog cranberry, and Labrador tea feel as stiff as museum-preserved specimens.

From the point I search the lake with the telescope, and later in the morning walk along much of the west and south shore. The tally is small: fifteen mallards, most of them drakes; twenty-five goldeneyes, most hens or immatures; forty-nine buffleheads, distributed in groups of two to eight each, which include both drakes and hens or immatures; three immature ruddy ducks together in a bay; one eared grebe; one common loon, grey and white; and a single coot. All the drakes are now restored to full nuptial colours.

I cross the short causeway to an island. A muskrat swims away from the edge quickly, then slows and meanders close to the shore. In the dense white spruce clothing the island's north slope four or five boreal chickadees move about, only occasionally showing themselves and lisping sweetly, calls quite unlike their usual husky *day-day-day*. They never leave the spruce, possibly because of the half dozen black-capped chickadees in the more open mixed-wood who tumble through the trees and shrubs, and often drop to the snow below the birches to pick up shed seeds. A golden-crowned kinglet seems to have attached himself to the group, although he does not come down from the trees. Unlike the ruby-crowned kinglet, this species has a vividly striped head. The eye is concealed in a black stripe; above is a strip of white, then black again, yellow, and finally a touch of orange. The rest of the plumage, however, is plain buffy-olive. In the tops of the birches siskins pick at the seed catkins. A male downy woodpecker flies in from somewhere and lands on a young aspen trunk. He hitches up and around for several feet, exploring, tapping lightly here and there. In little over a minute he flies away and soon out of sight.

I stop at the silverberry thicket. Many leaves have not yet fallen and are as rigid as leather that has lain forgotten in a barn or attic. The pale, ovoid fruits sheen in the sunlight, uniform like cut and polished stone jewelry.

Black-capped chickadees drift through the shrubs at the bog margin, while on the ridge above a flock of siskins has gathered in the birches. I walk along the ridge. Ahead a pallid form detaches from a tree, weaves away between the dusky trunks, dissolves like a wraith of fog. These northern great horned owls are often very pale and when seen in flight against a darker background may seem as white as a snowy owl.

The ridge and the forest end together in a sweeping, open slope. From the top I can see two ponds, one clearly just beyond the foot of the slope, the other screened by trees and shrubs. On the first is a drake mallard and several coots. A coyote moves through the sedge tussocks near the pond, slowly, pausing, turning this way and that, head low. I wonder if it is the same pup I watched vole hunting here two months ago, or the adult more recently. However, I stay where I am, removed.

Then I go down to the other pond which is much larger. Six goldeneyes swim and dive well out from shore; one is a black and white drake. Backed by tawny cattails, two drake mallards feed near the ice edge, flashing white bellies as they tip up.

I walk back, lingering where it is sunny to savour the glow. A few midges float in the air like small, tufted seeds. A spider crawls laboriously over the snow. It is very small with a slender red body and extremely long, very dark legs. A female hairy woodpecker advances up the corrugated bark of an old balsam poplar, not stopping to test the wood until she reaches the smooth thin bark of the crown. On the way she utters loud *pink*'s that strike the air like notes from a triangle, and at the end of these a short, thin chatter.

A solitary pine grosbeak moves south through the tree tops, perching often, each time whistling. Invariably a species to travel in company, this one seems lost and crying plaintively for his kind over the rolling barrens of the leafless forest. He is within sight and hearing for about two minutes, but the ache of his sadness endures in me long after.

Where the chickadees were on my way in, several juncoes flicker over the dark ground garnering energy for the long night ahead.

Fingers of shadow claw forward, lengthening, choking off the life of the sun, quenching all colours but the copper fires of the willows smouldering in hollows. In one a cow moose as black as coal browses untouched.

The ground is in ashes, but high above a sun-scored kestrel flings over the meadow like a boomerang, one whose return cannot be anticipated for months. The sun still seethes behind the trees, close to the horizon. I walk more quickly. A varying hare skitters away, mottled dun and white like the ground over which it runs and made visible only by movement.

An elk challenges the setting sun, the high, attenuated note piercing adversaries real or imagined, piercing the indifferent forest, piercing all ears.

November

The weather cooled yesterday before all the snow had melted. With overcast blinding the sun, the temperature dropped below freezing and snow squalls swept through all day in rapid succession.

This morning I expected a depth of snow. Instead there is only a skim and the clouds have vanished. The thermometer reads 10 degrees. I walk to the lakeshore. Glass-clear ice seals off all but the deepest parts. Concentrated, the late ducks, goldeneyes and mallards, blacken the water. Near shore a silent coot wanders baffled and disconsolate where yesterday it swam and dived.

Later in the day I visit other lakes, and several ponds. Only Astotin still has open water.

It stays below freezing for several days. The lake ice thickens and is whitened with snow. But the deep water stays open; the ducks do not leave.

One day, while the ice is still clear and I am counting the ducks through a telescope a mallard comes in to land. Mistaking ice for water he skids, sliding helplessly along with flailing wings until he flops into the real water. He preens and shakes for a long time before his composure is restored.

Today, November twelfth, the bald eagles arrive, four of them: a full adult with snowy head and tail, a bird probably in its third year with mottled brown and white head and dark-edged tail, and two young of the year who are almost as dark as ravens. They beat back and forth over the ducks, but none rise to the threat. An eagle is no match for a duck in full flight; its hope is to catch one taking off, or one weakened by a gunshot wound. This they do, although I have never witnessed the capture—just an eagle standing on the ice and pulling at a shapeless lump held down under its feet while two magpies try to dodge in and grab a piece, or flap irritatingly around the eagle's head.

Often the eagles perch in trees along the shore. The distant shape that might be an old nest or an odd arrangement of branches leaps to life in the telescope's circle. I am looking straight into a glittering, dark-pupiled eye made fierce by the heavy brow ridge above. The yellow beak is unbelievably massive and the head seems small by comparison. Except for the white, the plumage is a uniform sepia. Like beak and iris, the naked predatory feet are yellow, almost as yellow as the painted ones of museum mounts. The eagle falls out of the telescope and it is a few moments before I pick out its dark form over the ice, flying steadily.

For two days I bask in temperatures in the mid-forties. But ice does not melt, nor does the snow disappear. The ground is frozen and like pavement under the thin veneer of white.

The wind pours through the skeleton forest, then foams up and over the solid wall of spruce at the edge of the bog. Within, in the eternal autumn of the bog forest, the air is still, although I can hear the roaring turbulence careening on overhead. A squirrel's claws scratch as it hunches up a trunk. A downy woodpecker suddenly appears on a tree and taps here and there for a little while, changing position frequently. The spot of red on the back of his head seems very large and bright. Then he flies off, out of my circumscribed world. Boreal chickadees come close, whispering in my ears. Unseen, a grey jay flies into a tree above. Its querulous, hollow notes are engulfed by the wind and rushed away to another land.

The cold clamps down again bringing an inch of snow.

I stare across the lake, but am dazzled by the rising sun. Then the glare slowly diminishes and when I search now I can be sure—no dark gash of water; no ducks; no eagles. I feel a sharp pang of depression. Autumn is dead, winter born on the sixteenth of November.

I crunch through the shallow snow. Ahead is the gaiety of pine grosbeaks and I soon find the flock on a shrub slope picking the white berries of buckbrush. The bright males are like rose flowers scattered through the drab females and immatures. I count forty-five birds altogether, some in the shrubs, others resting or preening in the nearby aspen, although there is much coming and going between the two. Suddenly a tight little group of birds wheels up from the buckbrush into the trees, but keeps apart from the larger grosbeaks. In the bustle, colour and whistling of the grosbeaks I overlooked these Bohemian waxwings.

In the early afternoon at the edge of a meadow a white-tail buck lifts his head from browsing and turns toward me. His antlers curve gracefully, symmetrically tined, converging at the tips. In the sun his colouring is warm chestnut. Then he pivots, flags his tail and bounds into the forest. The meadow snow is laced with bison tracks and dotted with droppings. Grass has been bared by grazing and coarser clumps that were bypassed during the summer are now being eaten.

Later, in forest, a weasel runs along a fallen tree, stops to eye me with interest, drops down to the ground, pops up again, ripples to the end of the tree, looks back once and is gone. All but the top of its head is winter white.

December

The first few days of December are unseasonally mild as Pacific air floods over the frozen breadth of the plains. The sun shines, and some of the snow melts; bare ground is slimed with mud. A few small flies come out of hibernation.

Travel for the moment easy, I spend long hours on foot, yet now that most active life has withdrawn, the protracted time and distance bring relatively few rewards.

Pine grosbeaks, evening grosbeaks, and Bohemian waxwings are scattered in whistling and trilling flocks wherever buckbrush, snowberry, and bush-cranberries are concentrated. Pine siskins and the first redpolls briefly foliage the birches and cluster in the spruce. One day I hear a red-breasted nuthatch in a bog; on another find two ruby-crowned kinglets in a band of boreal chickadees. Magpies are constantly on the move in voluble groups of up to four, looking for anything edible—carrion, berries, or hazelnuts, an unwary mouse or vole. In dense forest blue jays replace the magpies. Of the three winter woodpeckers, I encounter the hairy most often, in proportion to its relative abundance. Many of the downies have migrated, while the pileateds require extensive tracts of mature forest and thus cannot be numerous in any place.

The red squirrel population is high this year, but not high enough to spill over significantly into winter poplar forest. Thus squirrel distribution matches that of white and black spruce. Whenever I enter stands of either I am verbally assaulted by the indignant owners.

Moose, elk, and deer, cryptically coloured, are difficult to see before they move, and then it is too late—they run far from the two-legged mammal they do not trust.

With almost predictable regularity coyotes wail their dusk choruses. Alike in tenor, yet infinitely variable in detail, I never hear them without listening. And no night is complete without the horned owl's bass hooting, usually in solo, occasionally duetted.

Cirrus clouds spin white gossamer across the sky. A halo blooms around the sun. The wind stirs and chills. Clouds thicken slowly, then mask the sun. Night closes in rapidly and by the next morning snow is falling, lacy flakes meshed. The white deepens, grasses and sedges are bowed down and covered. By the end of daylight the snow is four inches thick where the swirling wind has not drifted it. Still the flakes fall heavily on into the darkness, swaddling the earth in soft folds, shielding it from the bitterness of the future.

The sun rises gloriously into skeins of burning clouds. Then it is free. Crystals sparkle on the snow; shadows lie gently, lightly on its surface; all angles have been smoothed to curves. Nowhere can I see a single track and I am reluctant to make the first, to violate the pristine mantle. For a few days the transformation of the landscape inspires awe and wonder. The haunting reminders of summer are gone beyond recall.

❧ ❧ ❧ ❧ ❧ ❧ ❧

The eve of the solstice; the eve of a new year. I feel that something dramatic should happen to mark the event. But the river of time seems stilled in the breathless air. The day is clear and bright. The temperature creeps up to –20 degrees, no higher.

Black-capped chickadees have retreated to islands of spruce where red squirrels have retired to temporary dormancy. Two bull moose, one with huge antlers, the other with the spikes of a two-year old, browse compatibly among willows. In the afternoon I watch a bull bison standing broadside to the sun, beard almost touching the snow, eyes half closed, basking. Later I find magpie craters in a poplar grove. Alighting on the snow the birds scatter it with their beaks, and then rummage about in the leaves, scratching and pecking for food. Grouse tracks walk in a long, straight line into willow thickets. I move on and two grouse whirr up above the shrubs, then pitch down into poplar forest.

Silence departs from silence as a great horned owl floats away from a spruce into deeper gloom. Above, several red crossbills flutter about the cones, prying open the scales, and tweaking out the seeds. Fragments, and the birds' stuttering calls rain down on me. Farther on, a standing dead spruce surrounded by living holds a female black-backed, three-toed woodpecker. Attracted by the loosening bark, she works up the trunk slowly and methodically, flaking off chips by sliding her beak under and jerking her head sideways, and leaving behind a bright trail of unweathered wood. She is alone in the spruce, yet she frequently calls, a dry *pik* less forceful than the similar note of the hairy woodpecker.

Clarified and heightened by the cold the coyotes' sunset chorus pours through the dusk and soars into the last light of the zenith.

Night reaches over the sky. Trees blacken. The snow dims, its day-glare white softened. Stars begin to glitter in the east.

Appendix 1
Mammal checklist

Scientific names follow those in *The Mammals of Alberta* (1964) by Dr. J. Dewey Soper, except that for the most part subspecies are not recognized.

MOLES AND SHREWS (Insectivora)

Shrews (Soricidae)
saddle-backed shrew
(*Sorex arcticus*)
rarely seen; in bogs, as well as mixed-wood and poplar forest in damp depressions.
cinereous shrew
(*Sorex cinereus*)
infrequently seen; in a wide variety of forested and brushy habitats except very dry.

BATS (Chiroptera)

Insect-eating Bats (Vespertilionidae)
little brown bat
(*Myotis lucifugus*)
fairly common in small numbers; feeds in open at edge of forest; roosts in tree hollows and dense spruce foliage.

PIKAS, HARES AND RABBITS (Lagomorpha)

Hares and Rabbits (Leporidae)
snowshoe (varying) hare
(*Lepus americanus*)
uncommon; in a wide variety of forested, and occasionally shrubby habitats.
white-tailed jack rabbit
(*Lepus townsendii*)
rare; mainly open grassland in southeast section of park.

GNAWING MAMMALS (Rodentia)

Marmots and Squirrels (Sciuridae)
woodchuck (groundhog)
(*Marmota monax*)
rare; grass meadows with aspen groves and shrub islands.
Richardson ground squirrel ("gopher") (*Citellus richardsonii*)
locally common; in open grassland and parkland.
striped ground squirrel
(*Citellus tridecemlineatus*)
rare; parkland, and grassy lake and pond margins.
Franklin ground squirrel
(*Citellus franklinii*)
probably rare; parkland and shrub grass areas.
northern (least) chipmunk
(*Eutamias minimus*)
uncommon; generally in mature poplar forest and edge of same where ground is well-drained, often hilly.
red squirrel
(*Tamiasciurus hudsonicus*)
common; in white spruce stands, bogs and mixedwood; rare in poplar forest.
flying squirrel
(*Glaucomys sabrinus*)
rarely seen, as is nocturnal; inhabits mature mixedwood and white spruce stands, occasionally mature poplar.

Pocket Gophers (Geomyidae)
Richardson pocket gopher ("mole")*
(*Thomomys talpoides talpoides*)
quite common; edges of meadows, shrub slopes, and occasionally poplar forest where soils light and well-drained.

*although the pocket gopher is often called a "mole" it should not be confused with true moles which are insectivores (not rodents), closely related to the shrews, but do not occur in Alberta.

Beavers (Castoridae)
beaver
(*Castor canadensis*)
abundant; all aquatic habitats, creates and enlarges ponds by damming outlets.
Native Rats and Mice (Cricetidae)
white-footed mouse (deermouse)
(*Peromyscus maniculatus*)
abundant; all wooded habitats.
red-backed vole (mouse)
(*Clethrionomys gapperi*)
abundant; all wooded habitats.
meadow vole
(*Microtus pennsylvanicus*)
abundant; tall grass mixed with shrubs on uplands as well as borders of ponds, and sedge/willow fens.
little upland vole
(*Pedomys ochrogaster minor*)
may be locally common, colonial; parkland and shrubby habitats, well-drained.
muskrat
(*Ondatra zibethicus*)
abundant; aquatic; ponds and lakes.
Jumping Mice (Zapodidae)
jumping mouse
(*Zapus princeps*)
apparently uncommon; long grass and sedge at edges of forest and ponds.
American Porcupines (Erethizontidae)
porcupine
(*Erethizon dorsatum*)
uncommon, but may be increasing somewhat recently; mainly mixedwood and bog edges at least in winter where favoured food is white birch bark.

FLESH-EATING MAMMALS (Carnivora)

Wolves and Foxes (Candidae)
coyote
(*Canis latrans*)
usually common; all habitats within the park.
Bears (Ursidae)
black bear
(*Euarctos americanus*)
very rare; sightings were fairly frequent until 1914, then none until the summer of 1975 when one was seen briefly near the north boundary; it was later shot on a neighbouring farm.
Weasels, Otters (Mustelidae)
short-tailed weasel (ermine)
(*Mustela erminea*)
common; all habitats, but mainly forested.
least weasel
(*Mustela rixosa*)
rare; all habitats, but mainly forested.
long-tailed weasel
(*Mustela frenata*)
uncommon; predominantly parkland, shrubby and grassland habitats.
mink
(*Mustela vison*)
common; semi-aquatic, pond and lake margins even in winter.

American badger
(*Taxidea taxus*)
very rare; grassland and parkland.
striped skunk
(*Mephitis mephitis*)
uncommon, but may be increasing; mainly upland forest/meadow edges, parkland.
Cats (Felidae)
lynx
(*Lynx canadensis*)
uncommon; apparently not resident in the park; all habitats with shrub or forest cover.

EVEN-TOED HOOFED MAMMALS
(Artiodactyla)

Deer (Cervidae)
Manitoba elk (wapiti)
(*Cervus canadensis manitobensis*)
Population now maintained at about 350 head; a distinct subspecies of the interior plains.
mule deer
(*Odocoileus hemionus*)
very rare; population has declined steadily since 1950 for reasons not well understood.
white-tailed deer (white-tail)
(*Odocoileus virginianus*)
common, more so in the Isolation Area (south of hwy. 16) where there are very few elk; sporadic light population control.
moose
(*Alces alces*)
very common; population now maintained at about 350 head.
Cattle, Goats and Sheep (Bovidae)
plains bison (buffalo)
(*Bison bison bison*)
very common; population now maintained at about 400 head, in north section of park only.

wood bison (buffalo)
(*Bison bison athabascae*)
a small herd is maintained in the Isolation Area only.

Appendix 2
Bird checklist

This list has been updated to 1978. The names follow the standardized A.O.U. (American Ornithologists Union) *Check-list of North America* as revised in its 32nd supplement to the 1957 edition. However, alternative widely used common names, and ones only recently changed, are given bracketed. Scientific names have been omitted because of the standardized common names.

Abbreviations for each species status in Elk Island National Park:
R—year round resident (nests)
N—summer resident only (nests)
SV—summer visitor (does not nest)
M—migrant (usually spring & fall)
WV—winter visitor

LOONS

common loon (N)

GREBES

red-necked grebe (N)
horned grebe (N)
eared grebe (N)
western grebe (M)
pied-billed grebe (N)

PELICANS

white pelican (SV, M—rare)

CORMORANTS

double-crested cormorant (SV, M—rare)

HERONS, BITTERNS

great blue heron (N)
black-crowned night heron (N)
American bittern (N—rare)

SWANS, GEESE, DUCKS

whistling swan (M)
Canada goose (N, M)
white-fronted goose (M)
snow goose (M)
Ross' goose (M—rare)
mallard (N)
gadwall (N—rare)
pintail (N)
green-winged teal (N)
blue-winged teal (N)
cinnamon teal (N—rare)
American widgeon (baldpate) (N)
shoveler (N—rare)
wood duck (SV, M—very rare)
redhead (N)
ring-necked duck (N)
canvasback (N)
lesser scaup (bluebill) (N)
common goldeneye (whistler) (N)
Barrow's goldeneye (M)
bufflehead (N)
white-winged scoter (N)
surf scoter (SV, N?)
ruddy duck (N)
hooded merganser (M—very rare)
red-breasted merganser (M)
common merganser (M)

HAWKS, EAGLES, HARRIERS

goshawk (W)
sharp-shinned hawk (N)
Cooper's hawk (N)
red-tailed hawk (N)
broad-winged hawk (N)
Swainson's hawk (M, N?—rare)
rough-legged hawk (M)
golden eagle (M, W—rare)
bald eagle (M)
marsh hawk(M, N—rare)
osprey (M—rare)
peregrine falcon (M—very rare)
prairie falcon (FM—very rare)
merlin (pigeon hawk) (N)
American kestrel (sparrow hawk) (N)

GROUSE

spruce grouse (R—very rare)
ruffed grouse (R)
sharp-tailed grouse (R—rare)

PHEASANTS, PARTRIDGES

ring-necked pheasant (R—rare)
Hungarian partridge (R—rare)

CRANES

sandhill crane (M)

RAILS, COOTS

sora (N)
yellow rail (?)
American coot (N)

PLOVERS

semipalmated plover (M)
killdeer (N)
American golden plover (M)
black-bellied plover (M)

SNIPES, SANDPIPERS

common snipe (N)
spotted sandpiper (N)
solitary sandpiper (N)
willet (M—rare)
greater yellowlegs (M)
lesser yellowlegs (N)
pectoral sandpiper (M—rare)
Baird's sandpiper (M—rare)
least sandpiper (M—rare)
short-billed dowitcher (M)
stilt sandpiper (M—rare)
semipalmated sandpiper (M—rare)
western sandpiper (M—rare)
marbled godwit (M—rare)
sanderling (M—rare)

AVOCETS AND STILTS

American avocet (SV—rare)

PHALAROPES

Wilson's phalarope (N?—rare)
northern phalarope (M—rare)

GULLS, TERNS

herring gull (M, SV)
California gull (M)
ring-billed gull (SV)
Franklin's gull (SV)
Bonaparte's gull (M, SV)
Forster's tern (SV—rare)
common tern (SV)
black tern (N)

DOVES

mourning dove (N)

OWLS

great horned owl (R)
snowy owl (W—rare)
hawk owl (M, W—rare)
great gray owl (W—rare)
long-eared owl (N)
short-eared owl (?)
boreal owl (?)
saw-whet owl (R—rare)

GOATSUCKERS

common nighthawk (N)

HUMMINGBIRDS

ruby-throated hummingbird (N)

KINGFISHERS

belted kingfisher (M)

WOODPECKERS

yellow-shafted flicker (N)
pileated woodpecker (R)
yellow-bellied sapsucker (N)
hairy woodpecker (R)
downy woodpecker (R, N)
black-backed three-toed woodpecker (W—rare)
northern three-toed woodpecker (W—rare)

FLYCATCHERS

eastern kingbird (N)
great crested flycatcher (SV—rare)
eastern phoebe (N)
Say's phoebe (M—rare)
yellow-bellied flycatcher (M)
alder flycatcher (N)
least flycatcher (N)
western wood peewee (N)
olive-sided flycatcher (N—rare)

LARKS

horned lark (M)

SWALLOWS

tree swallow (N)
bank swallow (M—rare)
barn swallow (N)
cliff swallow (M—rare)
purple martin (N—rare, one colony)

CROWS, MAGPIES, JAYS

gray jay (R—rare)
blue jay (R)
black-billed magpie (R)
common raven (W— rare)
common crow (N)

TITMICE

black-capped chickadee (R)
boreal chickadee (R—rare)

NUTHATCHES

white-breasted nuthatch (M, N)
red-breasted nuthatch (M, N)

CREEPERS

brown creeper (M, W, N?—rare)

WRENS

house wren (N)
long-billed marsh wren (N)

MOCKINGBIRDS & THRASHERS

catbird (N)

THRUSHES, SOLITAIRES

American robin (N)
varied thrush (M—rare)
hermit thrush (M—rare)
Swainson's thrush (N)
gray-cheeked thrush (M)
veery (N)
mountain bluebird (N)
Townsend's solitaire (M—rare)

KINGLETS

ruby-crowned kinglet (M, N?)
golden-crowned kinglet (M, W)

PIPITS

water pipit (M)

WAXWINGS

Bohemian waxwing (W)
cedar waxwing (N)

SHRIKES

northern shrike (W—rare)
loggerhead shrike (N—rare)

STARLINGS

starling (N—rare)

VIREO

solitary vireo (M, N?)
red-eyed vireo (N)
Philadelphia vireo (N—rare)
warbling vireo (N)

WOOD WARBLERS

black-and-white warbler (N)
Tennessee warbler (N)
orange-crowned warbler (M, N?)
yellow warbler (N)
magnolia warbler (M)
Cape May warbler (M—rare)
yellow-rumped (myrtle) warbler (N)
black-throated green warbler (M)
bay-breasted warbler (M—rare)
blackpoll warbler (M)
pine warbler (M—very rare)
palm warbler (M, N—rare)
ovenbird (N—rare)
northern waterthrush (M)
Connecticut warbler (N)
mourning warbler (N)
common yellowthroat (N)
Wilson's warbler (M)
American redstart (N)

WEAVER FINCHES

English sparrow (R—rare)

MEADOWLARKS, BLACKBIRDS, ORIOLES

western meadowlark (N—rare)
yellow-headed blackbird (SV, N—rare)
red-winged blackbird (N)
northern (Baltimore) oriole (N)
rusty blackbird (M)
Brewer's blackbird (N)
common grackle (N)
brown-headed cowbird (N)

TANAGERS

western tanager (M, N?)

GROSBEAKS, BUNTING, FINCHES, SPARROWS

rose-breasted grosbeak (N)
evening grosbeak (W, M)
purple finch (N)
pine grosbeak (W, M)
gray-crowned rosy finch (W—very rare)
hoary redpoll (W)
common redpoll (W)
pine siskin (N, M)
American goldfinch (N)
red crossbill (M, W)
white-winged crossbill (W—rare)
savannah sparrow (N)
LeConte's sparrow (N)
sharp-tailed sparrow (very rare)
vesper sparrow (N—rare)
dark-eyed (slate-colored) junco
(M, N—rare)
dark-eyed (Oregon) junco (M—rare)
tree sparrow (M)
chipping sparrow (N)
clay-colored sparrow (N)
Harris' sparrow (M—rare)
white-crowned sparrow (M)
white-throated sparrow (N)
fox sparrow (M—rare)
Lincoln's sparrow (M—rare)
swamp sparrow (N)
song sparrow (N)
Lapland longspur (M—rare)
Smith's longspur (M—rare)
snow bunting (W—rare)

Appendix 3
Reptile, amphibian, and fish checklist

REPTILES (Reptilia)

Snakes (Serpentes)
Colubrid Snakes (Colubridae)
plains garter snake
(*Thamnophis sirtalis*)
apparently very rare; to be expected in grassland and parkland.

AMPHIBIANS (Amphibia)

Salamanders (Caudata)
Mole Salamanders (Ambystomatidae)
tiger salamander
(*Ambystoma tigrinum*)
apparently rare; pond and lake edges in spring, then damp forest.

Frogs and toads (Anura)
True Toads (Bufonidae)
Dakota (plains) toad
(*Bufo hemiophrys*)
common; pond and lake edges in spring, then damp forest.
Tree Frogs (Hylidae)
boreal chorus frog
(*Pseudacris triseriata maculata*)
abundant; never far from ponds and lakes.
True Frogs (Ranidae)
wood frog
(*Rana sylvatica*)
very common; pond and lake edges in spring, then damp forest.

FISH (Pisces)

Pike (Esocidae)
pike
(*Exos lucius*)
ascended Astotin Creek and spawned in Astotin Lake every spring until creek blocked by dams.

Minnows (Cyprinidae)
fathead minnow
(*Pimephales promelas*)
first observed in 1972, appears to be increasing considerably.
Suckers (Catostomidae)
sucker
(probably *Catostomus commersoni*)
as for the pike, it formerly entered Astotin Lake to spawn.
Sticklebacks (Gasterosteidae)
brook stickleback
(*Culaea inconstans*)
abundant and has always been a permanent resident of Astotin Lake; temporary populations have been found in connected ponds.

Appendix 4a Terrestrial invertebrate checklist

It must be emphasized that this listing is very fragmentary and serves as little more than a general indication of some of the groups so far found in the Park. Scientific names are from various sources, mainly titles in the Additional Reading list.

CENTIPEDES (Chilopoda)

Soil Centipedes (Geophilomorpha)
abundant in poplar forest

SPIDERS, MITES, TICKS (Arachnida)

True Spiders (Labidognatha)
Cobweb Weavers (Theridiidae)
Sheetweb Weavers (Linyphiinae)
common to abundant in forest undergrowth shrubs, also willow/alder thickets
Orb Weavers (Araneidae)
(*Araneus* spp.)
abundant; poplar forest undergrowth shrubs especially, also open shrub areas
Funnel Weavers (Agelenidae)
common; grassland
Wolf Spiders (Lycosidae)
abundant; poplar forest and shrubland, on ground
Crab Spiders (Thisidae)
flower spider (*Misumena vatia*)
very common on various flowers
(*Xysticus* spp.)
Jumping Spiders (Salticidae)
(*Salticus* spp.)
common; poplar forest undergrowth

Pseudoscorpions (Pseudoscorpiones)
common; leaf litter mainly; minute size

Harvestmen (Opiliones)
Daddy-longlegs (Phalangiidae)
common; poplar forest undergrowth

Mites and Ticks (Acarina)
Velvet Mites (Trombidiidae)
common; larvae parasitic on insects, adults free-living, eat insect eggs.

Sarcoptid Mites (Sarcoptiformes)
mange mite (*Sarcoptes scabei*)
occasionally infest coyotes

Ticks (Ixodides)
winter tick (*Dermacentor albipictus*)
common on moose, less often on elk, deer, and bison

INSECTS (Insecta)

Springtails, Snow Fleas (Collembola)
abundant; mainly in or at edge of forest

Mayflies (Ephemeroptera)
Small Mayflies (Baetidae)
nymphs aquatic; short-lived adults aerial, common briefly after emergence

Dragonflies and Damselflies (Odonata)
Darner Dragonflies (Aeshnidae)
green darner (*Anax* sp.)
blue darner (*Aeshna* spp.)
Common Skimmer Dragonflies (Libellulidae)
Narrow-winged Damselflies (Coenagrionidae)
bluet (*Enallagma ebrium*; *E. boreale*)
most species in this order are abundant; nymphs aquatic; adults over ponds and lakes and adjacent uplands; predatory

Grasshoppers, Crickets (Orthoptera)
Short-horned Grasshoppers (Acrididae)
abundant in dry meadows, shrub/herb slopes, sandhills
Long-horned Grasshoppers (Tettigoniidae)
bush katydids (Phaneropterinae)
occur mainly in forest undergrowth

Stoneflies (Plecoptera)
Common Stoneflies (Perlidae)
nymphs aquatic, adults aerial

Earwigs (Dermaptera)
Common Earwigs (Forficulidae)
one collected in building

Thrips (Thysanoptera)
abundant; in flowers

True Bugs (Hemiptera)
Minute Pirate Bugs (Anthocoridae)
common; predatory
Leaf Bugs (Miridae)
abundant; suck plant fluids
Assassin Bugs (Reduviidae)
common; predatory
Lace Bugs (Tingidae)
common; suck tree/shrub foliage
Seed (Chinch) Bugs (Lygaeidae)
common; most herbivorous
Leaf-footed (Squash) Bugs (Coreidae)
common; some herbivorous, some predatory
Shore Bugs (Saldidae)
common; moist habitats and near water; predatory and scavenging
Stink Bugs (Pentatomidae)
common; predatory and herbivorous
Damsel Bugs (Nabidae)
common on shrubs and herbs; predatory
Broad-headed Bugs (Alydidae)
common; herbivorous

Hoppers, Aphids, Cicadas (Homoptera)
Treehoppers (Membracidae)
(*Ceresa bubalis*)
Froghoppers or Spittlebugs (Cercopidae)
Leafhoppers (Cicadellidae)
Aphids or Greenfly (Aphidae)
above three families usually abundant; suck plant juices
Woolly and Gall Aphids (Eriostomatidae)
poplar leaf gall aphid
(*Pemphigus populicaulis*)
gall at base of leaf
(*P. populi-venae*)
gall on mid-vein of leaf
woolly alder blight
(*Prociphilus tessellatus*)
often common on young twigs and leaf veins
Pine and Spruce Aphids (Chermidae)
some form galls

Lacewings (Neuroptera)
Green Lacewings (Chrysopidae)
(*Chrysopa* spp.)
abundant; shrubby forest edge
Brown Lacewings (Hemerobiidae)
common in forest

Beetles (Coleoptera)
Tiger Beetles (Cicindelidae)
sandy and open habitats
Ground Beetles (Carabidae)
on ground in various habitats
Carrion Beetles (Silphidae)
on carrion and moist, decaying plants
Rove Beetles (Staphylinidae)
many species; one large one associated with bison dung, most predatory
Soldier Beetles (Cantharidae)
adults common on flowers (eat pollen and nectar); larvae predatory
Fireflies (Lampyridae)
periodically numerous; adults conspicuous during June mating flights
Click Beetles (Elateridae)
common in deciduous forest
Metallic Wood-boring Beetles (Buprestidae)
(*Dicerca divaricata*)
adult bronze; larvae bore in deciduous trees, especially white birch.
Ladybird Beetles (Coccinellidae)
two-spotted ladybird
(*Adalia bipunctata*)
very common; shrubby and forested habitats
Blister Beetles (Meloidae)
common; often on flowers; adults herbivorous; larvae parasitic, usually on grasshopper eggs
Chafer Beetles (Scarabidae)
June beetles (*Phyllophaga* sp.)
periodically common in forest; fly at dusk and later
Long-horned Beetles (Cerambycidae)
common; larvae tunnel in dead and occasionally living wood
Leaf Beetles (Chrysomelidae)
flea beetles (Alticinae)
common; larvae feed on plant roots; adults eat leaves
Snout Beetles (Curculionidae)
common; general herbivores
Short-winged Mold Beetles (Pselaphidae)
common; small beetles in rotting wood, moss, also ant nests; feed on moulds

Caddisflies (Trichoptera)
Large Caddisflies (Phyganeidae)
(*Agrypnia* sp.)
lakes; larval case short pieces of plant stem
Northern Caddisflies (Limnephilidae)
lakes; larval case very variable
Molannids (Molannidae)
(*Molanna* sp.)
lakes; larval case sand
Long-horned Caddisflies (Leptoceridae)
(*Triaenodes* sp.)
lakes, larval case short pieces of plant stem
(*Oecetis* sp.)
lakes; larval case sand or plant fragments

Butterflies and Moths (Lepidoptera)
The following abbreviations are used in the comments for this order: e—early, m—mid, l—late, v—very.

True Butterflies (Papillionoidea)
Swallowtails (Papilionidae)
tiger swallowtail
(*Papilio glaucus*)
adults common m-May to July; larvae on poplar, willow
Sulphurs, Whites, Marbles (Pieridae)
checkered white (*Pieris protodice*)
adults l-April to e-October; larvae on mustards
mustard white (*Pieris napi*)
adults m-May to m-August; larvae on mustards
cabbage white (*Pieris rapae*)
adults v-common e-May to m-October; larvae on mustards; introduced species from Europe
alfalfa (*Colias eurytheme*)
adults v-common e-June to m-October; larvae on legumes

Blues, Coppers, Hairstreaks (Lycaenidae)

American copper (*Lycaena phlaeas*)
adults June to August; larvae on docks in moist, open areas

bronze copper (*Lycaena thoe*)
adults July to August; larvae on docks, knotweeds in moist areas

purplish copper (*Lycaena helloides*)
adults June to September; larvae on docks, knotweeds in moist areas

saepiolus (greenish) blue (*Plebejus saepiolus*)
adults May to e-July; larvae on legumes

western tailed blue (*Everes amyntula*)
adults l-May to m-July; larvae on legumes

silvery blue (*Glaucopsyche lygdanus*)
adults May to l-July; larvae on legumes

Brush-footed Butterflies (*Nymphalidae*)

white admiral (banded purple) (*Limenitis arthemis*)
adults m-April to m-October; larvae on poplar, birch, willow mainly

red admiral (*Vanessa atalanta*)
adults May to m-October; larvae on stinging nettle, fold leaves

painted lady (*Vanessa cardui*)
adults l-May to m-October, numbers fluctuate markedly; larvae on thistles and other composites

Compton tortoiseshell (*Nymphalis j-album*)
adults m-April to e-October; larvae on birch, poplar, willow

Milbert's tortoiseshell (*Nymphalis milberti*)
adults April to m-October; larvae on nettles, colonial

mourning cloak (*Nymphalis antiopa*)
adults April to October, larvae on poplar, willow

satyr anglewing (*Polygonia satyrus*)
adults April to October; larvae on nettles

gray comma (*Polysonia progne*)
adults April to m-October; larvae on currants, gooseberry

green comma (*Polysonia faunus*)
adults April to m-October; larvae on willow, alder, currant, gooseberry

tawny crescent (*Phyciodes batesii*)
adults m-June to e-August; larvae on asters

pearl crescent (*Phyciodes tharos*)
adults l-May to l-August; larvae on asters

meadow fritillary (*Boloria toddi*)
adults l-May to e-September; larvae on violets

silver-bordered fritillary (*Boloria selene*)
adults June to e-September; larvae on violets

silverspot fritillary (*Speyeria atlantis helena*)
adults m-June to l-August; larvae on violets

great-spangled fritillary (*Speyeria cybele*)
adults l-June to l-August; larvae on violets

Meadow Browns (Satyridae)

ringlet (*Coenonympha tullia*)
adults l-May to m-August; larvae on grasses

common wood nymph (*Cercyonis pergala*)
adults l-June to e-September; larvae on grasses

common alpine (*Erebia epipsodea*)
adults e-June to e-July; larvae on grasses

Large Moths (Macrolepidoptera)

Sphinx Moths (Sphingidae)

twin-spotted sphinx (*Smerinthus geminatus*)
adults June to August; larvae on cherry, birch, willow

gray sphinx (*Hemaris modesta*)
adults June to August; larvae on snowberry, viburnum

hummingbird moth (*Hemaris thysbe*)
adults June to August; larvae on snowberry, viburnum

bumblebee moth (*Hemaris diffinis*)
adults June to August; larvae on snowberry, viburnum

Silkworm Moths (Saturniidae)

Polyphemus moth (*Antheraea polyphemis*)
adults July to August mainly; larvae on birch and other tree foliage

Looper Moths (Geometridae)

moon looper (*Campaea perlata*)
adults often common in July

Thyatirid Moths (Thyratiridae)

(*Habrosyne scripta*)
uncommon; adults in July in bogs, spruce forest

Prominents (Notodontidae)

tentacled prominent (*Cerura* sp.)
adults e-summer; larvae on willows, aspen m- to l-summer

Tent Caterpillars (Lasiocampidae)

forest tent caterpillar (*Malacosma disstria*)
adults August to September; larvae colonial on cherry, poplar, etc. but no tent constructed, May to e-July

Ctenuchid Moths (Ctenuchidae)

Virginia ctenucha (*Ctenucha virginica*)
adults l-June to July, diurnal; larvae on grasses

Tiger Moths (Arctiidae)

banded woolly bear (*Isia isabella*)
adults e to m-summer; larvae l-summer to fall, feed on plantains, hibernate

virgo tiger moth (*Callarethia virgo*)
adults July mainly; larvae on various herbs

tortoiseshell tiger moth (*Parasemia plantaginis*)
adults June and July

spotted tiger moth (*Parasemia parthenos*)
adults June and July mainly

Cutworms, Underwings (Noctuidae)

ilia underwing (*Catocala ilia*)
adults June to e-August

white underwing
(*Catocala relicta*)
adults July to August mainly; larvae on poplar, willow
Forester Moths (Agaristidae)
spotted forester
(*Alypia langtoni*)
adults m-summer mainly, more or less diurnal
Pericopid Moths (Pericopidae)
broad-winged pericopid
(*Gnophaela latipennis*)
adults July to e-August, diurnal
Tussock Moths (Liparidae)
white-marked tussock
(*Hemerocampa leucostigma*)
adults l-summer mainly; larvae on various trees and shrubs, including willow

Small Moths (Microlepidoptera)
Clearwing Moths (Sesiidae)
adults e to m-summer, diurnal; larvae bore into stems, roots
Leaf-rollers (Tortricidae)
adults e to l-summer mainly; larvae on various trees, shrubs
Leaf-blotch Miners (Gracilariidae)
poplar miner (*Phyllocnistsis* sp.)
larvae on aspen leaves, July mainly
Casebearer Moths (Coleophoridae)
larvae on many shrub/herb leaves, mainly June, July

True Flies (Diptera)
Crane Flies (Tipulidae)
larvae of many aquatic or semi-aquatic; adults aerial
Mosquitoes (Culicidae)
(*Aedes vexans*) and several other species
larvae and pupae aquatic; adults aerial
Midges (Chironomidae)
(*Chironomus tentans*) and others
phantom midges
(*Chaoborus* spp.)
(*Tanytarsus* spp.)
all the above in this family have larvae and pupae aquatic; adults aerial
Punkies, No-see-ums
(Ceratopogonidae)
(*Bezzia* or *Probezzia* sp.)
numbers are probably small; species that bite humans do not seem to be present
Fungus Gnats (Mycetophilidae)
adults abundant in fall; larvae in fungi and decaying plants
Fever (March) Flies (Bibionidae)
adults abundant in spring; larvae in roots and decaying plants
Black Flies (Simuliidae)
no human-biting species seem to be present
Gall Midges (Cecidomyidae)
very small, common flies
Soldier Flies (Stratiomyidae)
large, wasp-like; adults frequent flowers
Horse Flies (Tabanidae)
horse flies (*Tabanus* spp.)
deer flies (*Chrysops* spp.)
adults of the above common June to e-August; females bite mammals, including man
Stiletto Flies (Therevidae)
adults predatory; resemble Robber Flies, but less common
Robber Flies (Asilidae)
adults predatory on other insects
Bee Flies (Bombyliidae)
larvae parasitic on solitary bees and wasps; adults feed at flowers
Dance Flies (Empididae)
often adults form swarms that move up and down
Long-legged Flies
(Dolichopodidae)
small predatory flies of marshy habitats
Hump-backed Flies (Phoridae)
minute flies, often near decaying vegetation
Small-headed Flies (Acroceridae)
larvae internal spider parasites
Big-headed Flies (Pipunculidae)
larvae parasitic on leafhoppers and planthoppers
Hover (Flower) Flies (Syrphidae)
rat-tailed maggot
(*Eristalis tenax*)
hornet-mimic
(*Temnostoma vespiforma*)
and other species
These are abundant throughout spring and summer; feed at flowers; larvae eat aphids, scavenge, or live in ant nests
Meadow Flies (Otitidae)
moist habitats; larvae herbivorous or in decaying plants
Picture-winged (Fruit) Flies
(Tephritidae)
wings often patterned; frequent flowers; larvae herbivorous
Black Scavenger Flies (Sepsidae)
small; adults and larvae in and near animal droppings/ decaying plants
Marsh Flies (Sciomyzidae)
medium; adults near water; larvae parasitic/predatory on snails
Leaf-mining Flies (Agromyzidae)
This is the only family of insects that has been thoroughly collected and identified within the Park, and 71 species in 14 genera resulted. The larvae mine at least 69 species of plants, mostly herbaceous. The list is obviously too long to give here, but a copy is on file with the Park Naturalist.
Shore Flies (Ephydridae)
adults near water, most very small, predatory and/or nectar feeders; larvae mainly in mud
Frit flies (Chloropidae)
adults in various habitats; larvae of many species develop in grasses
Root Maggots (Anthomyiidae)
Some larvae mine leaves, creating large blotches; adults often at flowers
Dung Flies (Scatophagidae)
adults frequent mammal dung, especially bovine; larvae develop in dung
House Flies (Muscidae)
variable family; larvae in dung and decaying vegetation
Tachinid Flies (Tachinidae)
many species; larvae parasitic on other insects
Blow-flies, Bluebottles
(Calliphoridae)
many species; larvae in carrion or dung

Wasps, Bees, Ants, etc.
(Hymenoptera)
Common Sawflies
(Tenthredinidae)
adults variable, medium-large; larvae

leaf feeders, miners, gall makers
Cimbicid Sawflies (Cimbicidae)
large, often stout; larvae leaf feeders, mainly willow, elm
Horntails (Siricidae)
wood wasp (*Tremex* spp.)
large, cylindrical body; larvae bore in tree trunks
Stem Sawflies (Cephidae)
medium, black; larvae bore in tree trunks
Braconids (Braconidae)
small-medium; larvae parasitic
Ichneumon Flies (Ichneumonidae)
larvae parasitic; at least 3,000 species in North America
Chalcids (Chalcidoidea)
tiny to small; larvae parasitic or herbivorous
Gall Wasps (Cynipidae)
spiny rose gall
(*Diplolepis bicolor*)
larvae develop in spherical galls on rose leaves
Cuckoo Wasps (Chrysididae)
medium, metallic blue or green; larvae parasitic or coexisting in other bee/wasp nests
True Ants (Formicidae)
carpenter ants
(*Camponotus* spp.)
and many others
common in mature forest, both poplar and spruce, also grass meadows
Yellowjacket Wasps, Hornets (Vespidae)
hornets (*Vespa* spp.)
large paper nests usually well up in trees
yellowjackets (*Vespula* spp.)
paper nests on ground, usually in forest
Sphecid Wasps (Sphecidae)
variable family including sand, aphid, mud dauber and thread-waisted wasps
Leafcutting Bees (Megachilidae)
make nests in decaying wood with cells of cut-out pieces of leaves; adults solitary
Sweat Bees (Halictidae)
often small and metallic, attracted to sweat; semi-colonial nesting
Bumblebees, Honey Bees (Apidae)
bumblebees (*Bombus ternarius*)
and others
medium to large, hairy; queens overwinter; nest at or below ground level, social
honey bee (*Apis mellifera*)
small numbers enter from hives on farms adjacent to the Park

Appendix 4b
Aquatic invertebrate checklist

This list is somewhat more detailed than that for the terrestrial invertebrates since several limnological studies have been conducted in Astotin Lake.

SPONGES (Porifera)

Fresh-water sponges (Spongillidae)
may be common; lakes and ponds

JELLYFISH, SEA ANEMONES, HYDRAS (Coelenterata)

Hydras (Hydridae)
common to abundant in all standing water

FLATWORMS (Platyhelminthes)

Free-living Flatworms (Turbellaria)
Planarians (Planariidae)
common to abundant in ponds and lakes

Tapeworms (Cestoda)
Diphyllobothriidae
(*Schistocephalus solidus*)
internal parasite of brook stickleback and various fish-eating birds

ROTIFERS, WHEEL ANIMACULES (Rotatoria)

(*Keratella* sp.)
planktonic in lakes

SEGMENTED WORMS (Annelida)

Aquatic Earthworms (Oligochaeta)
(*Chaetogaster limnaei*)
epizoic on aquatic snails

Leeches (Hirudinea)
(*Erpobdella punctata*)
feeds on aquatic invertebrates, fish and frogs
(*Glossiphonia complanata*)
feeds on various invertebrates, especially snails

(*Haemopsis* sp.)
feeds on invertebrates, also vertebrates
(*Macrobdella* sp.)
feeds on vertebrate blood
(*Nephelopsis obscura*)
feeds on vertebrates, scavenges
(*Placobdella ornata*)
(*Placobdella parasitica*)
free-living in summer, otherwise on fish, turtles, snails

JOINT-LEGGED ANIMALS
(Arthropoda)

Shrimps, Crayfish, Lobsters, Crabs
(Crustacea)

Water Fleas (Cladocera)
(*Acroperus* sp.)
(*Bosmina longirostris*)
(*Ceriodaphnia* sp.)
(*Chydorus* sp.)
(*Daphnia magna*)
these are mainly planktonic species

Copepods (Copepoda)
(*Cyclops* sp.)
(*Diaptomus* sp.)
(*Paracyclops* sp.)
these are mainly planktonic species

Ostracods (Ostracoda)
(*Candona* sp.)
(*Cyclocypris ampla*)
(*Cyclocypris ovum*)
(*Cypria* sp.)
(*Cypridiopsis* sp.)
(*Limnocythere* sp.)
these are mainly planktonic species
Fresh-water Shrimps, Sideswimmers, Scuds (Amphipoda)
(*Eylais* sp.)
(*Gammarus lacustris*)
(*Hyallela azteca*)
these are bottom dwellers and also swim in shallow water

Spiders, Mites, Ticks (Arachnida)
Water Mites (Hydrachnidae)
(*Hydrachna* spp.)
common to abundant in ponds and lakes, also fens and seeps

Insects (Insecta)
Included here are only those insects in which the adult as well as the immature stages are largely aquatic.

True Bugs (Hemiptera)
Water Boatmen (Corixidae)
(*Dasycorixa hybrida*)
(*Dasycorixa rawsoni*)
Backswimmers (Notonectidae)
(*Notonecta kirbyi*)
Waterstriders (Gerridae)
(*Gerris dissortis*)
Ripple Bugs (Veliidae)
(*Microvelia pulchella*)
all these Hemiptera are common to abundant mainly in lakes and larger ponds

Beetles (Coleoptera)
Crawling Water Beetles (Haliphidae)
(*Haliphus* sp.)
Predacious Diving Beetles (Dytiscidae)
(*Acilius* sp.)
(*Graphoderus* sp.)
(*Hydroporus* sp.)
(*Dytiscus flaviventris*)
(*Lacophilus* sp.)
(*Ilybius* sp.)
common to abundant in both permanent and temporary water bodies
Whirlygig Beetles (Gyrinidae)
very common; larvae live below surface; adults are predatory and active on water surface

SNAILS, CLAMS, SQUIDS (Mollusca)

Snails, Limpets (Gastropoda)
Pouch Snails (Physidae)
(*Aplexa hypnorum*)
(*Physa gyrina*)
Orb Snails (Planorbidae)
(*Gyraulus parvus*)
(*Helosoma trivolvis*)
Pond Snails (Lymnaeidae)
gastropods are abundant in both ponds and lakes; orb snails and pond snails are particularly numerous

Clams, Mussels (Pelecypoda)
Fresh-water Clams (Sphaeriidae)
fingernail clam (*Pisidium* sp.)
these live on pond and lake bottom sediments and may be more numerous than collecting indicates

Appendix 5
Plant checklist

This was originally compiled in January 1972 by the author. Several species have been added since then, up to 1978, but the total list is still not complete, especially for fungi, mosses, lichens, sedges, grasses, and willows. It is based on the Park herbarium and photograph files, as well as field observations and research reports, including G. H. Turner's "Plants of the Edmonton District of the Province of Alberta," *Can. Field-Nat.* 63 (1949): 1–28. Names of the fungi are taken from *Mushrooms of North America* by Orson K. Miller Jr. (New York: E. P. Dutton & Co., [1972]). Botanical names of ferns and all higher plants follow *Flora of Alberta* by E. H. Moss (University of Toronto Press, 1949).

Abbreviations used in the comments are as follows:

Distribution
l—local, *w*—widespread, *?*—uncertain

Abundance
a—abundant, *c*—common, *s*—scattered, *r*—rare

Habitat
DG—disturbed ground (trails, roadsides, wallows)
GM—grass meadows
LM—lake margins
MF—mixed forest (white spruce, white birch, poplar)
PF—poplar forest
PM—pond margins
SB—black spruce bogs
SF—white spruce forest
SS—shrub slopes
SWM—sedge/willow meadows (fens)

FUNGI (Eumycophyta)

Mushrooms (Basidiomycetes)
Amanitas (Amanitaceae)
fly agaric (*Amanita muscaria*)
c, mainly *PF*; July to August, poisonous
Hygrophors (Hygrophoraceae)
(*Hygrophorus* spp.)
c, in *SF*; July to September, some poisonous
Milk Mushrooms (Russulaceae)
russulas (*Russula* spp.)
very *c*, mainly *PF*; August to September, some poisonous
Tricholomes (Tricholomataceae)
marasmius (*Marasmius androsaceus*)
c, in *SF*: July to September
Ink Caps (Coprinaceae)
shaggy mane
(*Coprinus comatus*)
a, mainly *GM*; August to September
Coral Fungi (Clavariaceae)
fringed coral (*Clavulina cristata*)
c, mainly *PF*; July to September
Boletes (Boletaceae)
(*Leccinum insigne*)
c, *PF*; July to September
(*Boletus* spp.)
c, *PF*; July to August
Polypores, Brackets
(Polyporaceae)
aspen heart rot
(*Fomes igniarius*)
wa, *PF*, *MF*; perennial bracket, parasitic
(*Polyporus* spp.)
wc, *PF*, *SF*; perennial on dead wood

Stinkhorns, Puffballs
(Gastromycetes)
True Puffballs (Lycoperdales)
giant puffball
(*Calvatia gigantea*)
wl, moist swales *PF*, *GM*; August to September
small puffballs
(*Lycoperdon spp.)*
wc, wood and duff *PF*, *SF*; August to September

Jelly Fungi
(Heterobasidiomycetes)
orange ear
(*Dacrymyces palmatus*)
ls, logs in *PF*, *MF*; July to September

Sac Fungi (Ascomycetes)
poplar canker
(*Hypoxylon pruinatum*)
ws, black cankers on mature poplar trunks, parasitic
common morel
(*Morchella angusticeps*)
wc, on ground *PF*, *MF*; May to June mainly
scarlet cup
(*Sarcoscypha hyemalis*)
ws, on dead wood *PF*; late April to May

CLUBMOSSES, HORSETAILS, FERNS, SEED PLANTS (Tracheophyta)

Clubmosses, Quillworts
(Lycopsida)
Clubmosses (Lycopodiaceae)
stiff clubmoss
(*Lycopodium annotinum*)
r, in a few *SB*
tree clubmoss
(*Lycopodium obscurum*)
r, moist *PF*

Horsetails (Sphenopsida)
Horsetails (Equisetaceae)
common horsetail
(*Equisetum arvense*)
wa, *DG*, edges of *PF*
water horsetail (*Equisetum fluviatile*)
lc, *LM*, *PM*
(*Equisetum pratense*)
ls, *SF*
woodland horsetail
(*Equisetum sylvaticum*)
wc, moist *PF* and *MF*, *SB* margins

Ferns (Pteropsida)
Adder's-tongue Ferns
(Ophioglossaceae)
grape (rattlesnake) fern
(*Botrychium virginianum*)
r, *PF*
Common Ferns (Polypodiaceae)
crested shield fern (*Dryopteris cristata*)
r, collected "under alder" by Turner
narrow shield fern (*Dryopteris spinulosa*)
ls, mainly *SB* margins
oak fern (*Gymnocarpium dryopteris*)
lc, *MF*

Conifers (Gymnospermae)
Pines, Spruces, etc. (Pinaceae)
tamarack, larch
(*Larix laricina*)
lc, *SB*
white spruce (*Picea glauca*)
lc, *MF*, *SF* (old stands mainly in northern half of park, including islands in lakes; spreading slowly)
black spruce (*Picea mariana*)
la, *SB*
jack pine (*Pinus banksiana*)
r, two saplings seen (S shore Astotin Lake, and one *SB*)

Flowering Plants (Angiospermae)
Cattails (Typhaceae)
common cattail
(*Typha latifolia*)
wa, *LM*, *PM*
Burreeds (*Sparganiaceae*)
giant burreed
(*Sparganium eurycarpum*)
wc, *LM*, *PM*
Pondweeds (Najadaceae)
slender-leaved pondweed
(*Potamogeton filiformis*)
lc, Astotin Lake
flat-stalked pondweed
(*Potamogeton friesii*)
lc, Astotin Lake
sago pondweed
(*Potamogeton pectinatus*)
lc, Astotin Lake
(*Potamogeton pusillus*)
ls, Astotin Lake
clasping-leaf pondweed
(*Potamogeton richardsonii*)
wa, lakes and ponds
large-sheath pondweed
(*Potamogeton vaginatus*)
lc, Astotin and Oster Lakes
Arrowgrasses (Juncaginaceae)
arrowgrass
(*Triglochin maritima*)
l, *SB*
Water-plantains (Alismataceae)
arrowhead
(*Sagittaria cuneata*)
ls, *LM*
Grasses (Gramineae)
couch grass (*Agropyron repens*)
ls, *DG*
slender wheat grass
(*Agropyron trachycaulum*)
?
water foxtail
(*Alopecurus aequalis*)
wc, *LM*, *PM*
slough grass
(*Beckmannia syzigachne*)
wc, *LM*, *PM*, wet depressions
fringed brome (*Bromus ciliatus*)
wc, *PF*
awnless brome
(*Bromus inermis*)
ls, *DG*
bluejoint (*Calamagrostis canadensis*)
wa, *LM*, *PM*, *SWM*
drooping wood reed
(*Cinna latifolia*)
lc, especially moats around *SB*
hairy wild rye (*Elymus innovatus*)
?, collected on Sandhills
great manna grass
(*Glyceria grandis*)
wc, *LM*, *PM*, wet depressions
foxtail barley
(*Hordeum jubatum*)
lc, *DG* (mainly roadsides)
mountain rice grass
(*Oryzopsis asperifolia*)
wc, *PF*
reed canary grass
(*Phalaris arundinacea*)
lc, *LM*
timothy (*Phleum pratense*)
lc, *GM*
reed, giant reed grass
(*Phragmites communis*)
lc, shores and islands of Astotin Lake
annual bluegrass (*Poa annua*)
ws, *DG* (especially along animal trails)
plains bluegrass (*Poa interior*)
?, *GM*
fowl bluegrass (*Poa palustris*)
ws, *LM*, *PM*, wet depressions
alkali grass (*Puccinellia distans*)
ls, common only at Soapholes
spangletop
(*Scolochloa festucacea*)
wc, *LM* and PM
Sedges (Cyperaceae)
water sedge (*Carex aquatilis*)
?, collected in one *SB*
awned (marsh) sedge
(*Carex atherodes*)
wa, *LM*, *PM*, *SWM*
golden sedge (*Carex aurea*)
lc, near *PM*
(*Carex bebbii*)
?, collected in shallow pond at Sandhills
(*Carex brunnescens*)
ls, mainly "moats" around *SB*
(*Carex crawfordii*)
?, collected by Turner near Astotin Lake; also seen near Oster Lake
(*Carex foenea*)
?, collected near viewpoint on Lakeview Trail (Astotin Lake)
(*Carex pauperculus*)
lc? in *Sphagnum* of wet *SB*
greater beaked sedge
(*Carex* sp., near *rostrata/ rhynchophysa*)
l, collected in Ridge Pond on Lakeview Trail
beaked sedge
(*Carex rostrata*)
?, collected in Ridge Pond on Lakeview Trail
(*Carex stipata*)
ws, wet depressions
(*Carex trisperma*)
l, *SB* (collected by Turner)
spike rush
(*Eleocharis palustris*)
l (one collected); colony by long bridge at lake edge, Lakeview Trail
cotton grass (*Eriophorum vaginatum*)
lc, *SB*
wool grass (*Scirpus atrocinctus*)
lc, *LM*, *SWM*
small-fruited bulrush
(*Scirpus microcarpus*)
ls, *LM* (shallow water)
great bulrush (*Scirpus validus*)
lc, *LM* (shallow water)
Arums (Araceae)
sweet flag, calamus
(*Acorus calamus*)
r (one collected)
water arum, marsh calla
(*Calla palustris*)
c, in shallow ponds
Duckweeds (Lemnaceae)

common (lesser) duckweed
(*Lemna minor*)
wa, lakes and ponds
ivy duckweed (*Lemna trisulca*)
wa, lakes and ponds
Rushes (Juncaceae)
wire rush (*Juncus balticus*)
r?, collected near Astotin Lake
Lilies (Liliaceae)
nodding onion
(*Allium cernuum*)
ws, *PF*
fairy bells (*Disporum trachycarpum*)
wa, *PF*
western wood lily
(*Lilium philadelphicum*)
ws, *PF*, *SS* (most numerous on Sandhills)
lily-of-the-valley, mayflower
(*Maianthemum canadense*)
wa, *PF*, *MF*
star-flowered Solomon's-seal
(*Smilacina stellata*)
lc, *SS*, open *PF*
three-leaved Solomon's-seal
(*Smilacina trifolia*)
la, *SB*
white camas
(*Zygadenus elegans*)
r, roadside of Oster Trail (poisonous)
Irises (Iridaceae)
blue-eyed grass (*Sisyrinchium montanum*)
lc, *GM*
Orchids (Orchidaceae)
spotted coralroot
(*Corallorhiza maculata*)
lc, *PF*
pale coralroot
(*Corallorhiza trifida*)
?, (recorded by Turner)
northern green orchid
(*Habenaria hyperborea*)
ws, *PF*, *MF* (occasionally)
bracted green orchid
(*Habenaria viridis* var. *bracteata*)
s?, shrubby slopes of 1946 Burn
Willows (Salicaceae)
balsam poplar
(*Populus balsamifera*)
wa, *PF*, *MF*
aspen poplar
(*Populus tremuloides*)
wa, *PF*, *MF*
beaked willow
(*Salix bebbiana*)
wc, *LM*, *SWM*
pussy willow (*Salix discolor*)
wc, mainly *SWM*
dark sandbar willow
(*Salix melanopsis*)
ls, *LM* (Astotin Lake)
myrtle-leaved willow
(*Salix myrtillifolia*)
= *S. nova-angliae*)
ws, *LM*, *PM*
moose willow (*Salix petiolaris*)
wa, *SWM*, *LM*, *PM*
broad-leaved willow
(*Salix planifolia*)
?, collected on Elk Island, Astotin Lake
balsam willow
(*Salix pyrifolia*)
wc, *LM*, *PM*, *Sb* margins
autumn willow
(*Salix serissima*)
ls, *SWM*
Birches, Hazels (Betulaceae)
river alder (*Alnus tenuifolia*)
wc, edges of *SB*, *SWM*, *PM*
white birch (*Betula papyrifera*)
ws, on upland *PF* and *MF*, var. *neolaskana lc* in *SB*, *SWM*
swamp birch (*Betula pumila*)
ls, *SB*, *SWM*
beaked hazel (*Corylus cornuta*)
wa, in *PF*, *SS*, less common in *MF*
Hemps (Cannabinacceae)
hop (*Humulus lupulus*)
r, one plant known near Administration Building
Nettles (Urticaceae)
stinging nettle (*Urtica gracilis*)
wc, *PM*, *LM*, moist areas in *PF*, conspicuous on beaver lodges
Buckwheats (Polygonaceae)
striated knotweed
(*Polygonum achoreum*)
?, *DG*
water smartweed
(*Polygonum amphibium*)
lc, *LM*, *PM*
common knotweed
(*Polygonum aviculare*)
lc, *DG*
wild buckwheat
(*Polygonum convolvulus*)
lc?, *DG* (bison wallows and trails)
smartweed
(*Polygonum lapathifolium*)
lc, *DG*
sheep sorrel (*Rumex acetosella*)
ls?, *DG* (bison wallows and trails)
western dock
(*Rumex occidentalis*)
ws, *PM*, *SWM*, moist *DG*
Goosefoots (Chenopodiaceae)
Russian pigweed
(*Axyris amaranthoides*)
lc, *DG*
lamb's quarters
(*Chenopodium album*)
lc, *DG*
strawberry blight
(*Chenopodium capitatum*)
ls, *DG*
Pinks (Caryophyllaceae)
sandwort (*Arenaria lateriflora*)
wc, *PF*
mouse-ear chickweed
(*Cerastium nutans*)
lc, *DG*
white campion (*Lychnis alba*)
r, *DG* (roadsides)
(*Stellaria crassifolia*)
?, *LM*, *PM*
long-leaved chickweed
(*Stellaria longifolia*)
ws, moist *DG*, edges of *SWM*, *SB*
long-stalked chickweed
(*Stellaria longipes*)
lc, *GM*
Water-lilies (Nymphaeaceae)
yellow pond-lily
(*Nuphar variegatum*)
lc, in ponds
Hornworts (Ceratophyllaceae)
hornwort (*Ceratophyllum demersum*)
wa, in lakes and ponds
Crowfoots (Ranunculaceae)
baneberry (*Actaea rubra*)
wc, *PF* and *MF*; fruits poisonous (two color varieties—red fruited and white fruited)

Canada anemone
(*Anemone canadensis*)
wc, edges and open areas of *PF*
bank anemone (*Anemone riparia*)
ws, edges and open areas of *PF*
floating marsh marigold
(*Caltha natans*)
ls, shallow ponds
marsh marigold
(*Caltha palustris*)
wc, *PM*, *SWM*, wet areas in *PF*
purple clematis
(*Clematis verticellaris*)
r, slopes in *PF*
small-flowered buttercup
(*Ranunculus abortivus*)
lc, moist areas in *PF*
tall buttercup
(*Ranunculus acris*)
l, colony in moist ditch by Moss Lake Road at Moss Lake
white water crowfoot
(*Ranunculus aquatilis*)
r (one collected)
yellow water crowfoot
(*Ranunculus gmelinii*)
r (one collected)
Macoun's buttercup
(*Ranunculus macounii*)
wc, *PM*, *SWM*, moist *DG*
cursed crowfoot
(*Ranunculus sceleratus*)
lc, *LM* and *PM* (on mud)
veiny meadow rue
(*Thalictrum venulosum*)
wc, *PF*, *SS*, edges of *GM*

Fumitories (Fumariaceae)
golden corydalis
(*Corydalis aurea*)
ws, mainly on *DG*

Mustards (Cruciferae)
tower mustard (*Arabis glabra*)
?
shepherd's purse
(*Capsella bursa-pastoris*)
ls, *DG*
bitter cress
(*Cardamine pensylvanica*)
?
tansy mustard
(*Descurainia sophia*)
ls, DG
wormseed mustard
(*Erysimum cheiranthoides*)
?
common peppergrass
(*Lepidium densiflorum*)
lc, *DG*
yellow cress (*Rorippa islandica*)
?, collected in channel between Little Tawayik and Tawayik Lakes
pennycress (*Thlaspi arvense*)
lc, *DG*

Sundews (Droseraceae)
round-leaved sundew
(*Drosera rotundifolia*)
r, only known in one *SB*

Saxifrages (Saxifragaceae)
golden saxifrage
(*Chrysosplenium iowense*)
r, *LM* and intermittent streams
bishop's cap (*Mitella nuda*)
wc, *PF*, *MF*, *SF*
Grass-of-Parnassus, parnassia
(*Parnassia palustris*)
ws, *PM*, *SWM*, moist depressions
wild black currant
(*Ribes americanum*)
ws, moist *PF*, edge *SWM*
skunk currant
(*Ribes glandulosum*)
wc, edge *SWM*, intermittent stream courses, moist *PF*
northern black currant
(*Ribes hudsonianum*)
ws, intermittent stream courses
bristly black currant, swamp currant (*Ribes lacustre*)
r, *SF* (collected on Elk Island, Astotin Lake)
wild (northern) gooseberry
(*Ribes oxyacanthoides*)
wa, *PF*, *MF*
wild red currant (*Ribes triste*)
wa, *PF*, *MF*

Roses (Rosaceae)
agrimony (*Agrimonia striata*)
wc, open *PF*, edges of *GM*
saskatoon, serviceberry, Juneberry
(*Amelanchier alnifolia*)
ws, open *PF*, *SS*, edges of *GM*
hawthorn, round-leaved haw
(*Crataegus chrysocarpa*)
r, edges of *PF* and *GM*
woodland strawberry
(*Fragaria vesca*)
ls, *SF*, *MF*
wild strawberry (*Fragaria virginiana*)
wa, **GM**, open *PF* and edges, *SS*
yellow avens (*Geum allepicum*)
wc, open *PF*, edges of *GM*
large-leaved avens
(*Geum macrophyllum*)
ws, open *PF*
water avens, purple avens
(*Geum rivale*)
?
silverweed (*Potentilla anserina*)
lc, *GM*, Soapholes
graceful cinquefoil
(*Potentilla gracilis*)
ls, *GM*
rough cinquefoil
(*Potentilla norvegica*)
ws, *DG*
marsh cinquefoil
(*Potentilla palustris*)
wc, *SWM*, *PM*
pin cherry
(*Prunus pensylvanica*)
ls, *SS*, edges of *PF*
choke cherry, black choke cherry
(*Prunus virginiana* var. *melanocarpa*)
ws, *SS*, edges of *PF*
prickly rose (*Rosa acicularis*)
wa, *PF*, *MF*
dwarf raspberry
(*Rubus acaulis*)
lc, *PM*, *LM*, margins of *SB*
cloudberry
(*Rubus chamaemorus*)
la, *SB*
trailing raspberry, dewberry
(*Rubus pubescens*)
wc, *PF*, *MF*
wild raspberry
(*Rubus strigosus*)
wc, open *PF*

Peas (Leguminosae)
milk vetch
(*Astragalus agrestis*)
r, colony on roadside near Administration Building
tall milk vetch
(*Astragalus canadensis*)

r, sandy soil Moss Lake Trail
caragana (*Caragana arborescens*)
naturalized near residences and golf clubhouse
white peavine
(*Lathyrus ochroleucus*)
wa, *PF*
purple peavine
(*Lathyrus venosus*)
wa, *PF*
alfalfa (*Medicago sativa*)
r, naturalized on Elk Island, Astotin Lake
reflexed loco weed
(*Oxytropis deflexa*)
r, one colony known on *DG* at bottom of Fire Tower Hill (Astotin Lake)
golden bean
(*Thermopsis rhombifolia*)
r, photograph in Park collection
alsike clover
(*Trifolium hybridum*)
lc, *DG* (mainly roadsides)
red clover
(*Trifolium pratense*)
lc, *DG* (mainly roadsides)
white clover
(*Trifolium repens*)
lc, *DG*, *GM*
vetch (*Vicia americana*)
wc, *PF* and *SS*

Geraniums (Geraniaceae)
Bicknell's geranium
(*Geranium bicknellii*)
ls, *DG*

Touch-me-nots (Balsaminaceae)
touch-me-not, jewelweed
(*Impatiens capensis*)
lc, *LM*, *PM*, moist areas in *PF* and *MF*

Violets (Violaceae)
early blue violet (*Viola adunca*)
lc, *GM*
marsh violet (*Viola palustris*)
lc, margins of *SB*
western white violet
(*Viola rugolusa*)
wc, *PF*, *MF*
kidney-leaved violet
(*Viola renifolia*)
lc, damp *PF* and *MF*

Oleasters (Elaeagnaceae)
silverberry, wolf willow
(*Elaeagnus commutata*)
ls, slopes near Astotin Lake (Elk Island and N of Sandy Beach), scattered on slopes of 1946 Burn
buffaloberry
(*Shepherdia canadensis*)
wc, *PF*

Evening Primroses (Onagraceae)
fireweed
(*Epilobium angustifolium*)
wc, *DG*, open areas in *SB* and *PF*
marsh willow-herb
(*Epilobium glandulosum*)
wc, *PM* and *LM*
enchanter's nightshade
(*Circaea alpina*)
lc, open wet ground along temporary streams

Water-milfoils (Haloragidaceae)
water milfoil
(*Myriophyllum exalbescens*)
wa, in lakes and ponds

Ginsengs (Araliaceae)
wild sarsaparilla
(*Aralia nudicaulis*)
wa, *PF* and *MF*

Carrots (Umbelliferae)
caraway (*Carum carvi*)
r, one collection near main road, also near residences
water hemlock
(*Cicuta bulbifera*)
ls, *PM*, *SWM*
broad-leaved water hemlock
(*Cicuta douglasii*)
ls, *LM*, *SWM*
cow parsnip
(*Heracleum lanatum*)
wc, open *PF*, edge *SWM*
sweet cicely
(*Osmorhiza depauperata*)
ls, *PF*
snake-root
(*Sanicula marilandica*)
wa, *PF*, *SS*, edge *GM*
water parsnip (*Sium suave*)
wc, *LM*, *PM*, *SWM*
meadow parsnip (*Zizia aptera*)
r, *GM*

Dogwoods (Cornaceae)
bunchberry
(*Cornus canadensis*)
wa, *PF*, *MF*, *SF*
red-osier dogwood
(*Cornus stolonifera*)
ws, *PM*, *LM*, *SWM*, moist *PF*

Wintergreens (Pyrolaceae)
pink wintergreen
(*Pyrola asarifolia*)
wa, *PF*, *MF*, *SF*
one-sided wintergreen
(*Pyrola secunda*)
ws, *PF*, *MF*
white wintergreen
(*Pyrola elliptica*)
ls, *PF* and *MF*

Heaths (Ericaceae)
bog rosemary
(*Andromeda polifolia*)
ls, *SB*
bearberry
(*Arctostaphylos uva-ursi*)
lc, on Sandhills
Labrador tea
(*Ledum groenlandicum*)
la, *SB*
small bog cranberry
(*Oxycoccus microcarpus*)
ls, *SB*
dwarf bilberry
(*Vaccinium caespitosum*)
r (one collected)
blueberry
(*Vaccinium myrtilloides*)
lc, *SB*, Sandhills
bog cranberry
(*Vaccinium vitis-idaea*)
lc, *SB*

Primroses (Primulaceae)
fairy candelabra
(*Androsace septentrionalis*)
lc, *GM*
fringed loosestrife
(*Lysimachia ciliata*)
ws, *PF*
tufted loosestrife
(*Lysimachia thyrsiflora*)
ls, *SWM*
Gentians (Gentianaceae)
felwort, northern gentian
(*Gentianella amarella*)
wc, *GM*, open *PF*
spurred gentian
(*Halenia deflexa*)
ws, open *PF*
buckbean
(*Menyanthes trifoliata*)
lr, *SB* (only if open water present)
Dogbanes (Apocynaceae)
spreading dogbane
(*Apocynum androsaemifolium*)
wc, *PF*
Phloxes (Polemoniaceae)
collomia (*Collomia linearis*)
lc, *DG* (especially sandy soil)
Borages (Boraginaceae)
stickseed (*Hackelia americana*)
lc, *SF* and *DG* on Elk Island, Astotin Lake
bluebell, lungwort
(*Mertensia paniculata*)
wc, *PF*, *MF*
Mints (Labiatae)
giant hyssop
(*Agastache foeniculum*)
ws, *SS*, *PF*
hemp nettle
(*Galeopsis tetrahit*)
ls, *DG*
water horehound
(*Lycopus asper*)
ls, *LM* (on mud)
wild mint (*Mentha arvensis*)
wc, *LM*, *PM*, *SWM*
skullcap
(*Scutellaria galericulata*)
wc, *LM*, *PM*, *SWM*
hedge nettle (*Stachys palustris*)
wc, *LM*, *PM*, *SWM*, moist areas in *PF*

Figworts (Scrophulariaceae)
red (common) paintbrush
(*Castilleja miniata*)
ls, open *PF*
yellow toadflax
(*Linaria vulgaris*)
r, *DG*
owl clover (*Orthocarpus luteus*)
r, *GM* (moist, sandy ground)
slender blue beard-tongue
(*Penstemon procerus*)
r, on Sandhills
American brooklime (speedwell)
(*Veronica americana*)
lc, *SWM*, margins of *SB*, moist areas in *PF*
hairy speedwell
(*Veronica peregrina*)
?, *DG*
Bladderworts (Lentibulariaceae)
common bladderwort
(*Utricularia vulgaris*)
lc, in ponds and sheltered *LM*
Plantains (Plantaginaceae)
great (common) plantain
(*Plantago major*)
lc, *DG*
Madders (Rubiaceae)
northern bedstraw
(*Galium boreale*)
wa, *PF*, *MF*, *SS*, edge *GM*
small bedstraw
(*Galium trifidum*)
lc, *SWM*, *PM*, *LM*
sweet-scented bedstraw
(*Galium triflorum*)
ls, *PF*, *MF*
Honeysuckles (Caprifoliaceae)
twinflower (*Linnaea borealis*)
wc, *PF*, *MF*, *SF*
twining honeysuckle
(*Lonicera dioica*)
wc, *PF*
bracted honeysuckle,
black twinberry
(*Lonicera involucrata*)
wc, *PF*, *MF*, edge *SWM*
snowberry
(*Symphoricarpos albus*)
wc, *PF*, *MF*
buckbrush, western snowberry
(*Symphoricarpos occidentalis*)
wa, *GM*, *SS*, open *PF* and *MF*

low-bush cranberry
(*Viburnum edule*)
wa, *PF*, *MF*, *SF*
high-bush cranberry
(*Viburnum trilobum*)
ws, *PF*
Moschatel (Adoxaceae)
moschatel
(*Adoxa moschatellina*)
lc, damp *PF* near streams, ponds and bogs where other undergrowth reduced
Teasels (Dipsaceae)
blue buttons
(*Knautia arvensis*)
lc, colony along *DG* of west fireguard (Isolation Area)
Harebells (Campanulaceae)
harebell
(*Campanula rotundifolia*)
lc, *GM* (abundant in Sandhills)
Sunflowers, Composites
yarrow milfoil
(*Achillea millefolium*)
wc, *GM*, *SS*
Siberian yarrow
(*Achillea sibirica*)
wc, *DG*, open *PF*
false dandelion
(*Agoseris glauca*)
r, probably only Sandhills
woodland pussytoes
(*Antennaria neglecta*)
lc, *MF*, *SF*, *PF*
pussy-toes
(*Antennaria nitida/parviflora*)
lc, *GM*
plains (leafy) arnica
(*Arnica chamissonis*)
lc, *GM*, open *PF*
heart-leaved arnica
(*Arnica cordifolia*)
r, one seen on Oster Trail
biennial sagewort (wormwood)
(*Artemisia biennis*)
ls, *DG*
dragonwort
(*Artemisia dracunculus*)
ls, *DG*
prairie sagewort (wormwood)
(*Artemisia ludoviciana*)
lc, *GM*, *DG*

Lindley's aster (*Aster ciliolatus*)
wa, *PF*, *MF*
showy aster (*Aster conspicuus*)
wa, *PF*, *MF*
white prairie aster
(*Aster pansus*)
r, *GM*, *SS*
willow aster (*Aster hesperius*)
wc, *LM*, *PM*
rush-leaved aster
(*Aster junciformis*)
l, in *SB*
smooth aster (*Aster laevis*)
lc, *DG*, *SS*, open *PF*, Sandhills
hairy aster (*Aster modestus*)
lc, *SWM*
purple-stemmed aster
(*Aster puniceus*)
wc, *LM*, *PM*, *SWM*, moist areas in *PF* and *MF*
nodding (smooth) beggar-tick
(*Bidens cernua*)
lc, *LM* and *PM* (on mud)
ox-eye daisy (*Chrysanthemum leucanthemum*)
r, *GM* (near Administration Building)
creeping thistle, Canada thistle
(*Cirsium arvense*)
ls, *DG*
Flodman's thistle
(*Cirsium flodmanii*)
ls, *SS*
soft-spined thistle
(*Cirsium foliosum*)
ls, *GM*
annual hawksbeard
(*Crepis tectorum*)
wc, *DG*
Canada fleabane
(*Erigeron canadensis*)
ls, *DG*
smooth (meadow) fleabane
(*Erigeron glabellus*)
ws, *GM*; *c*, Sandhills
hirsute fleabane
(*Erigeron lonchophyllus*)
ls, *DG*
Philadelphia fleabane
(*Erigeron philadelphicus*)
lc, *LM*, *PM*, *SWM*
daisy fleabane
(*Erigeron strigosus*)
ls, *DG*
low cudweed
(*Gnaphalium uliginosum*)
ls, occasional on open mud
gumweed (*Grindelia squarrosa*)
ls, only found at Soapholes
rhombic-leaved sunflower
(*Helianthus laetiflorus*)
r, *GM* near east boundary
Canada hawkweed
(*Hieracium canadense*)
wc, open *PF*, *SS*
blue lettuce (*Lactuca pulchella*)
ls, *DG*
scentless chamomile
(*Matricaria maritima*)
r, *DG*
pineapple weed
(*Matricaria matricarioides*)
lc, *DG*
palmate coltsfoot
(*Petasites palmatus*)
wc, *PF*
arrow-leaved (marsh) coltsfoot
(*Petasites sagittatus*)
wa, *LM*, *PM*, *SWM*
vine-leaved coltsfoot
(*Petasites vitifolius*)
ls, moist areas in *PF* and *MF*
Petasites plants of hybrid origin also occur frequently.
black-eyed Susan
(*Rudbeckia serotina*)
r, *DG*
marsh ragwort
(*Senecio congestus*)
lc, *LM* and *PM* (on mud)
tall ragwort,
cut-leaved groundsel
(*Senecio eremophilus*)
ls, *SWM*, edge *PF*
balsam groundsel
(*Senecio pauperculus*)
r, *SWM* edges, moist *MF*
common groundsel
(*Senecio vulgaris*)
r, *DG* around buildings mainly
mountain goldenrod
(*Solidago decumbens*)
ls, *SS*, sandy *GM*
tall smooth goldenrod
(*Solidago gigantea*)
wa, *SS*, open *PF*, edges *GM*
common goldenrod
(*Solidago lepida*)
wa, *SS*, open *PF*, edges *GM*, *PF*

Glossary

anther, see *stamen*

axil, the angle at the base of a leaf stem where it joins the twig.

bolete, of the family Boletaceae; mushroom-like fungi which have minute pores instead of gills.

carunculated, covered with lumpy knobs.

chitin, the horny outer covering that makes up the exoskeleton of arthropods (animals with jointed body and legs such as, insects, some crustaceans, and spiders).

chloropid fly, of the family Chloropidae; small grey or yellow-and-black flies most of whose larvae feed in grass stems. Adults of some species are nectar feeders.

coleophorid moth, of the family Coleophoridae; tiny moths whose larvae make external cases and feed by mining leaves.

corolla, the petals of a flower taken together as a unit.

crustacean, of the class Crustacea; largely aquatic animals and including lobsters, shrimp, crabs, barnacles, water fleas, and sow bugs (wood lice).

dolichopodid fly, of the family Dolichopodidae; small to medium flies with long legs and often metallic coloured bodies. They frequent meadows and marshes where the adults are predatory or scavenge small insects floating on water.

cynipid wasp, of the family Cynipidae; very small wasps whose larvae make or live in galls.

elytra, the hard forewings, characteristic of beetles, protecting the membranous hind wings which are used for flight.

epidermis, the outer layers of the skin of an animal; also the thin surface layer of plants, especially herbaceous kinds.

feldspar, a crystalline mineral made up of aluminum silicates in combination with sodium and/or potassium; colour varies from white to pink, grey or bluish; present in almost all igneous (granite-like) rocks.

femur, the first long bone (thigh) of the hind limb; in arthropods, the first long segment of any leg.

fen, an area of low, marshy ground dominated by grasses, sedges, and rushes.

fuscous, a brownish-grey.

gabbra, a blackish coarsely crystalline igneous rock containing much iron and magnesium.

gracilariid moth, of the family Gracilariidae; adults are very small, but the larvae make conspicuous mines in leaves; the poplar miner is often very common on aspen (see Appendix 4a).

hydrostatic organ, paired internal sacs of phantom midge larvae (*Chaoborus* spp.) which allow it to rise or sink in the water as the gas volume is increased or decreased.

hymenopteron, of the order Hymenoptera; which includes bees, wasps, ants, ichneumons, sawflies, and others (see Appendix 4a).

ichneumon fly, of the family Ichneumonidae; parasitic, slender-bodied hymenoptera often possessing a long, thread-like ovipositor.

incisor, cutting teeth at the front of the jaw, between the fang-like canines; rodents, hares, and rabbits lack canines but have much enlarged incisors.

interference, refraction and reflection of light waves by the physical structure of feathers that results in a coloured band of the spectrum being seen, somewhat as in a rainbow.

invertebrate, any animal without an internal skeleton and backbone; includes jellyfish, worms, molluscs, crustaceans, insects, and the like.

keratinous, the variably flexible, often coloured, non-cellular albuminoid forming horns, hoofs, nails, hair, scales, and feathers.

legume, of the family Leguminosae; commonly called the pea family, and comprising mainly herbaceous plants.

liverwort, a small green plant resembling a moss, but usually with larger and fewer leaves; or with a flat, lobed and leathery form; most grow on damp to wet ground, wood, or rock.

mollusc, of the class Mollusca; invertebrates that include snails, slugs, clams, squids, and octopi.

muscid fly, of the family Muscidae; which includes the common house fly, but also many other species not associating with man.

ochre, a brownish-yellow or light tan colour.

ovipositor, a specialized organ at the tip of the abdomen of many female insects for depositing eggs.

parabolic, a more or less bowl-shaped curve.

pedicel, a slender stalk supporting a single flower or leaf.

perianth, the petals and sepals of a flower taken together, and used when both these parts resemble each other closely in colour and shape, as in lilies and irises.

pistil, the reproductive part of a flower consisting of an ovary and a slender style supporting the enlarged, sticky stigma which receives pollen.

proboscis, the extended mouth parts of some invertebrates, especially insects, which are adapted for piercing and/or sucking.

pointillism, a style of painting in which the paint is applied in dots of pure colour rather than linear brush or palette-knife strokes.

raceme, a long flowerhead with a single, vertical axis which bears many short-stemmed flowers.

rhizome, a perennial, more or less horizontal underground root-like stem which sends up leafy shoots at intervals, and from which roots grow down into the soil.

rufous, a bright, red-brown.

saprophytic, applied to plants and referring to those obtaining nutrients from dead or decaying organic material.

sepal, the outer bracts of a flower that are usually green and leaf-like, and protects the petals when the flower is in bud.

sepsid fly, of the family Sepsidae; small, slender black or purple flies frequently found at moist decaying plants and animal dung.

spiracle, a breathing pore in the body wall of insects which connects with the tracheal system (air tubes).

stamen, the reproductive part of a flower consisting of a filament and an enlarged anther at the tip where pollen is produced.

strobili (sing. strobilus), cone-like structures on clubmosses and horsetails which produce spores; may also refer to the cone-like reproductive spikes of pines and related conifers.

thermal, a rising column of warm air which usually forms over localized areas where the ground surface is hotter than on adjacent sites.

thorax, the middle of the three major divisions of an insect's body i.e. head, thorax, and abdomen; wings and legs are on the thorax.

tomentose, on plants, any part that is covered with dense white hairs.

tortricid moth, of the family Tortricidae; adults of most species are small; larvae roll leaves or bind several together with silk while they feed, and often remain to pupate.

umber, a moderate, slightly yellowish-brown.

umbo, in fungi refers to a raised area at the centre of a mushroom cap.

vermiculated, marked with irregular, closely spaced, fine lines.

Additional reading

Borror, D. J. and R. E. White. *A Field Guide to the Insects of America North of Mexico.* Boston, Massachusetts: Houghton Mifflin Co. 1970. 404 pp., col., b&w illus.

Cormack, R. G. H. *Wild Flowers of Alberta.* Edmonton, Alberta: Hurtig Publishers. 1977. 415 pp., col. illus.

Hardy, W. G. (editor-in-chief). *Alberta: A Natural History.* Edmonton, Alberta: MisMat Corporation Ltd. 1967. 343 pp., col. illus.

Hodge, R. P. *Amphibians and Reptiles in Alaska, the Yukon and Northwest Territories.* Alaska: Northwest Publishing Company. 1976. 78+ pp., col. illus., maps.

Hooper, R. R. *Butterflies of Saskatchewan: A Field Guide.* Regina, Saskatchewan: Department of Natural Resources, Saskatchewan. 1973. 216 pp., illus., some col.

Klots, E. B. *The New Field Book of Freshwater Life.* New York: G. P. Putnam's Sons. 1966. 372+ pp. col., b&w illus.

Lang, A. H. *Guide to the Geology of Elk Island National Park: The Origin of its Hills and Other Scenery.* Ottawa, Ontario: Geological Survey of Canada. Misc. Rept. No. 22. 1974. 30 pp., illus., map.

Levi, H. W. and L. R. Levi. *A Guide to Spiders and their Kin.* New York: Golden Press. 1968. 160 pp., col. illus.

Miller, O. K. Jr. *Mushrooms of North America.* New York: E. P. Dutton & Co., Inc. [1972.] 360 pp., col., b&w illus.

Mitchell, R. T. and H. S. Zim. *Butterflies and Moths: A Guide to the More Common American Species.* New York: Golden Press. 1964. 160 pp., col. illus., maps.

Murie, O. J. *A Field Guide to Animal Tracks.* Boston, Massachusetts: Houghton Mifflin Co. 1954. 374 pp., b&w illus.

Salt, W. R. and J. R. Salt. *Birds of Alberta.* Edmonton, Alberta: Hurtig Publishers. 1976. 480+ pp., col., b&w illus., maps.

Schalkwyk, H. M. E. *Mushrooms of the Edmonton Area; Edible and Poisonous.* Edmonton, Alberta. 1975. 31 pp., b&w illus.

Paetz, M. J. and J. S. Nelson. *The Fishes of Alberta.* Edmonton, Alberta: Government of Alberta. 1970. 282 pp., col. ill., maps.

Shuttleworth, F. S. and H. S. Zim. *Nonflowering Plants: Algae, Fungi, Lichens, Mosses, Ferns.* New York: Golden Press. 1967. 160 pp., col. illus.

Soper, J. D. *Mammals of Alberta.* Edmonton, Alberta: The Hamly Press Ltd. 1964. 402+ pp., col., b&w illus., maps.

Stebbins, R. C. *A Field Guide to Western Reptiles and Amphibians.* Boston, Massachusetts: Houghton Mifflin Co. 1966. 279 pp. col., b&w illus., maps.

Vance, F. R., J. R. Jowsey and J. S. McLean. *Wildflowers Across the Prairies.* Saskatoon, Saskatchewan: Western Producer Prairie Books. 1977. 194+ pp., col. illus.

Index